Understanding Aircraft Composite Construction

Basics of Materials and Techniques for the Non-engineer

by Zeke Smith

Aeronaut Press, Napa, California

Understanding Aircraft Composite Construction

Basics of Materials and Techniques for the Non-engineer

By Zeke Smith

Published by:
Aeronaut Press
1076 Stonebridge Dr.
Napa, CA 94558-5345 USA

Order from:
Pathway Book Service
Lower Village
Gilsum, NH 03448   USA
Tel 800/345-6665, Fax 603/357-2073

Publisher's Cataloging in Publication
(*Prepared by Quality Books, Inc.*)

Smith, Zeke.
Understanding aircraft composite construction : basics of materials and techniques for the non-engineer / by Zeke Smith
p. cm.
Includes bibliographical references and index.
LCCN: 94-96418
ISBN 0-9642828-1-X

1. Airplanes—Materials. 2. Composite materials.   I. Title

TL699.C57S65 1996   629.134
                                                QBI96-20161

Author's Note:

This book is intended to make the amateur pilot/builder of small Experimental catagory aircraft feel confident and comfortable about the use of modern composite materials, techniques, and structures.

The intended audience is essentially the readers of *Sport Aviation*, *Kitplanes* and similar magazines.

Design and material concepts are explained with a bare minimum of mathmatics.

It is not a design manual; the detailed engineering of safe aircraft components is beyond the scope of this book. The reader will understand why composite aircraft are designed the way they are and how they differ from aircraft designed in other materials.

This book discusses three general areas:

(1) Basic theory of strong materials, structures, and composites

(2) Actual materials and structures used in small aircraft

(3) Construction techniques used by the home builder and by the prefabricated kit manufacturers

The homebuilt aircraft community is united by many things: the love of personal flight, fascination with the endless subtlety of aircraft, and the sharing of advancing knowledge and experience among friends. As a builder, pilot, and writer, I have tried to bring a part of that knowledge and experience into this book. I know that there is much more that others, more qualified than I, can add to the progress of Experimental Aviation. Their comments and corrections are welcome.

Zeke Smith
Napa, California
March, 1996
psmith@napanet.net

Acknowledgements:

Many fellow pilot/builders have given me freely of their time, friendship, and knowledge, in hangars and in airport coffee shops. Some said, "You're a writer? Do a book!" I don't remember all their names, but I hope that they will remember me and will accept my thanks now.

EAA Chapter 338, San Jose, CA, got me started in experimental aviation. Thanks, guys, for showing me all that stuff.

Rex Taylor, then of Viking Aircraft, helped me with my early Task Research Dragonfly.

Special thanks to Rich Trickel, of Tri-R Technologies and Lance Niebaur and staff at Neico Aviation, for patiently showing me their state-of-the-art aircraft and processes.

Mike Hoke and Greg Kress of Abaris Training, Reno, were generous with their help.

Dr. Pat Martin and (future) Dr. Dave West read the theoretical portion of the work to keep me right. Dan Porter read it as a professional pilot. Dick Bibb, old friend and long-time pilot, read it because he cared.

Jeanette West and Dr. Brenda Bryant read much of the manuscript to see if it made sense and remedied what didn't.

My wife Jean supported the project from the beginning and understood why I was so intense about it. Thanks, love.

Zeke Smith

... *A special remembrance to the late Bernhardt Hurwood, lifelong writer, optimist, and encourager.   It took too long, Bern.*

The Author

Zeke Smith is a free-lance technical writer. He wrote technical manuals for high-tech Silicon Valley clients for eleven years, following thirty years as a Senior Engineer in the IBM development laboratories in San Jose, CA. He taught electrical engineering for one year at San Jose State University.

He has been an active pilot since 1981. He is a past president of EAA Chapter 338, San Jose, and is now active in Chapter 167, Napa, CA. He owns and flies a Beech *Sundowner*.

ntroduction

This book will introduce the non-engineer pilot to the general principles and techniques used in current composite aircraft construction. Four groups of readers will find this book important:

- Those seriously interested in constructing a homebuilt aircraft from one of the new designs, who wish to consider a composite design, and need a basis for an informed decision among composite, metal, wood, or other construction material.

- Those who have chosen to build a composite aircraft and wish to gain a more fundamental understanding of the design principles so that the best design selection can be made.

- Those who are building a composite aircraft or repairing an existing Experimental Category aircraft and who wish to make modifications, however minor, that may not be described in the construction manual.

- Those experienced builders and users of composite aircraft construction techniques who wish to design a new aircraft and will consider including in the new design some of the methods and techniques described here.

We homebuilders owe a debt of gratitude to Burt Rutan and his crew for his many innovations, his imagination, his get-things-done attitude, and his engineering integrity. His *Vari-Eze* and *Long-EZ* cast their shadows over nearly everything in homebuilt composite construction. He led the way in using revolutionary new materials in new ways to create new airplanes that influenced the many designs that followed. Now the homebuilt aircraft community has a much larger and richer family of designs, materials, and techniques to choose from, many not yet widely understood in the General Aviation community.

This book will discuss the science of basic materials that supports our understanding of all composite materials: why things break, why other things are strong and how it is that we can make strong structures from fragile fibers and soft resins.

The book will then briefly cover the basics of the mechanics of structures as used in aircraft: how forces build up in a simple beam (like a wing spar) and where the forces go. This understanding will be extended to more complex structures.

Next, the book will discuss specific fiber and matrix systems actually in current use and available to the home builder and the basic techniques of handling, storage, and use.

Specific methods of mold making, mold design, and the production of molded parts will be shown. These techniques greatly speed up the completion of a project and assure a quality result.

Last, methods will be demonstrated for assembling, reinforcing, attaching, aligning, and finishing composite structures.

To repeat: *this book is for the non-engineer pilot*, the regular member of the local EAA chapter. It's kept simple. There is no heavy-duty math because the concepts as explained here don't require it.

Today's composites make practical the construction and ownership of excellent small personal aircraft for the thousands who lack the skill, patience, or money to build an aircraft with older technologies. This book will show how it works.

Understanding Aircraft Composite Construction

Table Of Contents

Chapter 4 -
Theory of the Structural Sandwich - - 4-1

Chapter 5 -

Chapter 10 -
Basic Moldless Wet Layup Techniques - 10-1

Chapter 13 -
Using Molds - - - - - - - - 13-1

Chapter 1
Strong Materials

Basic Materials Principles

Intuition About Materials

We can't get through life if we don't have some intuitive ideas about the strength of materials. We don't cross a stream on a log if the log looks rotten, but we may cross the same stream on a smaller log if the log looks fresh and green. We know about how strong a shoelace is. Children eventually learn how not to break everything they touch. We buy rope for an application in a size that seems to be "about right" in our experience, knowing that steel cable is "stronger" than cotton rope and that nylon is somewhere in between. We know about how thick a wooden shelf should be. We put strapping tape on a package so that it seems safe, but we do not *engineer* these daily decisions.

In this chapter, we will briefly discuss the elementary principles that explain how strong materials work, how they differ, and why we have strong materials at all. In later chapters, we will look at how forces exist in practical structures.

As pilots and airplane builders, we probably have a better intuitive feel for materials than the general population, based partly on actual experience with a variety of aircraft materials. The FAA allows us to make simple repairs on certificated aircraft, generally following handbooks of standard practice (some very old) for the material and structure in question. Pilots are usually a conservative lot (not a bad thing for drivers of flying machines!) and they tend to stick to the familiar. A pilot trained in a *J-3 Cub* or a *Citabria*, may choose to build a tube and fabric aircraft (with its many, many pieces). If the pilot was trained in a *Cessna 152* or owns a *Bonanza*, she may elect to build an aluminum plane. Some just like wood. The author regularly visits with active builders at neighboring airports and is surprised at how many have never considered a composite project. If asked why not, the answer is usually something like "Well, I never was really comfortable with 'plastic' since a friend hit a rock with his Corvette and tore up a fender. Besides my plastic garbage can has a crack in it, so I don't want to use that stuff in my airplane."

The new combinations of high performance materials we call *composites* are generally outside the experience of most pilots and many designers. These materials are very different from wood, steel, and aluminum. Often, the designs and construction techniques are radically different from familiar practice. It is only to be expected that a pilot would be cautious.

This book will give the cautious pilot and builder some understanding of the basic principles and techniques that govern the design of composite structures. To do this, it is necessary to review some of the basic principles behind any strong material, including how and why structures made of practical materials fail.

Action and Reaction

When undergraduate engineers take a course in "Statics", they learn that if an object is stationary, it must be because all the forces on that object balance, or *add to zero*. For example, we can hang a simple hardware store spring scale from a tree, put a one pound fish on it, and observe that the scale pointer moves one inch and stops. The weight of the fish is balanced by the reaction force of the deflected spring. A parked airplane may press on the pavement with a weight of 1,500 pounds applied through the three wheels. All the component weights, including the engine, wing structure, battery, fuel tanks, and whatever, are brought through the structure in a complex way, but they must add up to the 1,500 total pounds measured at the wheels. But the airplane does not move, so the pavement must press up at each of the three wheels exactly enough to balance the total load.

What actually happens is that the concrete under the tires deflects a tiny bit, exactly as the spring scale does, so that the necessary opposite force is generated. Concrete pavement has the useful property of being very stiff, so that we do not *see* this tiny deflection. If we put three precision spring scales under the wheels (as we would do for a center of gravity measurement) then the parked plane would find a new rest position, and we would read the required deflection on the three scales. *The point is that every solid material, even diamond, must deflect under load.*

Forces in a Solid

In any solid, the atoms are held in a rigid position by the electrical forces between them. Atomic forces are the basis for chemical reactions between atoms and we need not distinguish, for this purpose, between "chemical" and "electrical" bonds. The strength of the bonds varies widely from one solid material to another. When a load is placed on a solid, the solid must deflect just enough to develop an equal counter force. If the atoms can't change position, then the only movement in the system must be the bonds between the atoms. Therefore, the stiffness of a material is determined by the strength of these bonds.

A very simple way to think of the nature of a solid is as an array of tiny solid balls, each ball connected to its neighbors by springs. (A coil spring mattress would be a good analogy.)

Under compression, the interatomic *springs* between the layers move together, as when a person sits down on a mattress, and a net upward force is produced which is equal and opposite to the force applied. Similarly, under tension, the *springs* are stretched and a resisting force is generated in the opposite direction.

These tiny atomic deflections in the body of a solid have been confirmed by actual experiments. If a single-crystal specimen is stretched in a testing machine by a small percentage, the atomic spacing will be found to vary in exactly the same percentage.

Stress and Strain

The words "stress" and "strain" have been grossly misused in ordinary conversation to mean many things other than the simple meaning intended by materials scientists. For a materials scientist, *stress* only means *load per unit area,* so:

$$\text{stress} = \text{load / unit area} = \sigma = P/A$$
$$(\text{where } P = \text{load, } A = \text{area})$$

We see from an aircraft parts catalog that an "AN" aircraft bolt family made from 8740 alloy steel is rated at 125,000 pounds per square inch. This is a statement of allowed stress in tension *before* safety factors are applied. The AN3 bolt from this family is 3/16 inch in diameter and has an area of 0.0276 in^2, so it will carry 3,451 pounds. The corresponding AN8 1/2 inch bolt has an area of 0.196 in^2 so will take 24,500 pounds. The *loads* are different, but the *stress* is the same. The two examples look the same to the bonded iron atoms inside the bolts because the atomic deflections necessary to carry the load are the same. The designer must provide a sufficiently large bolt to carry the required load at the allowed maximum *stress.*

Strain is defined by materials scientists as the *deformation per unit length* resulting from a stress, so:

$$\text{Strain} = \text{deformation/length} = \varepsilon = d/L$$
$$(\text{where } d = \text{deformation; } L = \text{original length})$$

If we had a very long bolt, say 100 inches long, and loaded it so that it stretched (deformed) 1/2 inch to 100.5 inches, the strain would be .5/100, or 0.005, or 0.5%. Strain is always just a fraction or ratio, so has no unit dimensions.

Stress and strain are expressions of how the atoms of a structure are deformed by a load and are independent of the size and shape of the structure. Stress and strain are characteristics of the <u>material</u>, not the structure.

Hooke's Law

It may seem obvious, after going through the above argument, that twice the deformation of a body should produce twice the reaction force, but it wasn't obvious for a very long time. People didn't understand that ordinary bodies had to deform to resist a load. Robert Hooke (1639-1713) was the English physicist who studied springs in clocks and first stated that the reaction force of a spring was proportional to the force on the spring. This simple statement is known now as *Hooke's law*.

In practical engineering materials, Hooke's law is valid for *elastic* solids. An elastic solid is one which returns to its original shape when a load is removed. Rubber, glass, steel, and diamond are elastic. (Grease, in contrast, is an *inelastic* solid.)

Ordinary elastic materials obey Hooke's Law up to the load where the specimen breaks or the material flows.

Actually, Hooke's Law is only an approximation valid for the small deformations we would allow in safe engineering design. Experiments with very strong silicon "whiskers" (more about whiskers later) show that the interatomic force diminishes for very large strains which we do not achieve in practice. (The specimen breaks first.)

Young's Modulus

Engineers who followed Hooke didn't seem entirely clear on the difference between elasticity as an intrinsic property of a material and as a property of the shape and dimensions of the particular object they were deforming. The Corvette automobile has used composite wheel suspension springs instead of the classical steel springs. The composite spring was *designed differently* from the corresponding steel spring, but the two designs had the *equivalent* function. As we'll see later, we can make different spring designs from different materials which *act* the same, i.e., have the same deformation under load.

Thomas Young (1773-1829) understood (in 1807) that it was important to characterize the intrinsic "stiffness" of a material independent of its application. He recognized that, for useful materials:

stress/strain = σ/ε = constant = E

And the constant E is the very important parameter of a material called Young's modulus, or the "stiffness". We will make frequent use of this concept in airplane design, where we need to know how far a structure will bend under a given load.

Strain (ε) is just a ratio, a plain number, so Young's modulus has the *same dimensions as a stress*, or force per unit area. Numerically, it is the stress which produces 100% strain, so E is typically a rather large number, which does not represent a practical stress. (Rubber can stand 100% strain, but steel can't!)

Values of E for some typical materials are shown on the next page.

Note that Hooke's Law, and therefore Young's Modulus, applies equally for both tension and compression. Remember that the model is simply that of pulling or pushing on the interatomic bond "springs". Some practical materials, like concrete and cast iron, contain internal flaws which open up with tension loads and propagate cracks, so these materials will test much weaker in tension than in compression. (But we hardly ever use concrete or cast iron in our aircraft designs.)

Strength

Strength is the *stress* required to break a structure. Strength may be measured for either the tension or compression case, but is usually quoted for tension. (Concrete, of course, is an exception because concrete is generally assumed to have zero tensile strength.)

Strength makes no statement about stiffness [Young's modulus (E)]. Porcelain is very stiff (high E), but so brittle that it has low

Young's Modulus for Typical Materials

Material	Young's Modulus (E) (1,000s of psi)
Rubber	1
Cheap bottle "plastic"	200
Composite matrix resins (typical)	800
Nylon fiber	800
Bamboo lignin (the matrix)	1,000
Wood (approx., spruce)	2,000
Plaster of Paris	2,000
Concrete	2,500
Bamboo fiber	4,000
Bone	6,000
Magnesium metal	6,000
Portland cement	10,000
Ordinary glasses (beer)	10,000
Aluminum (2024T3)	10,600
Kevlar 49 fiber	19,000
Steel	30,000
Carbon (graphite) fiber	34,000
Aluminum oxide (sapphire)	60,000
Diamond	170,000

strength and breaks easily. Nylon is flexible (low E), but very strong (high breaking stress).

Some typical tensile strengths for common and aircraft materials are given below, in thousands of pounds per square inch. These figures came from different sources, for tests under different conditions, and can only be used as a rough guide, still there are probably some surprises here!

Material	*Breaking Strength 1,000s psi stress*
Iron & Steels:	
Cast iron	10-40
Hardware store mild	60
Automotive sheet (high)	200-300
Piano wire (brittle)	450
Other Metals:	
Aluminum, cast (pure)	10
Magnesium alloys	30-40
Aluminum (alloys)	20-80
Titanium alloys	100-200
Non-metals:	
Wood, spruce, across grain	0.5
Wood, spruce, along grain	15
Glass, (beer, window)	5-25
Porcelain, good quality electrical	50
Fibers (Isolated - not in laminate):	
Hemp rope	12
Spider web	35
Silk	50
Cotton	50
Flax	100
Fiberglass (E-glass)	500
Kevlar 49	700

Thus, Strength is an expression of breaking stress, not the total load required to break a structure.

Strength and Structure:

Our simple model of a solid consisting of tiny atomic balls held to-gether by electrical "springs" is generally satisfactory only for very small deflections. Practical materials of practical size actually fail for other reasons long before the interatomic bonds are broken. Practical materials are usually complex and have an internal structure which affects the measured strength.

A *composite* material is a combination of a strong material (usually as a fiber) imbedded in a weaker matrix. The combination produces a new material with properties which can greatly outperform the properties of either component alone.

Concrete and Wood As Structures

Concrete actually meets our definition of a *composite* material, so when we test a specimen of concrete, we test a complex structure consisting a mixture of cement, sand, and granite pebbles, which work together to carry the load. In architecture, designers have been able to do great things while always keeping the materials in compression. Roman arches, domes of great churches, and but-tressed cathedrals are all examples of successful designs which use materials only in compression.

Wood is primarily a complex *structure* of cellulose. Under a micro-scope, wood is seen to consist of an array of tubular closed cells sur-rounding empty space. Depending on the state of seasoning, the spaces may contain some free water. The table of strengths above shows radically different figures for wood for strength tests along and across the grain. See from the table that cotton (which is essen-tially pure cellulose fibers) tests in tension very much stronger than practical wood and approaches the strength figure for mild steel. Wood, therefore, as a *structure* of cellulose, does not achieve the tensile strength of *pure* cellulose. As we shall see, wood is a surprisingly efficient material, due to its combination of low density, stiffness, and toughness. It is subject to detrimental effects such as rot and swelling, however.

Theoretical Strength

For a very long time, engineers have wondered why is steel "stronger" than copper, why it is that diamond is so hard and rubber so soft, or why don't all solid materials have the same strength - or none? So far in this discussion, we have used the model of the tiny atomic balls attached to each other with springs. It would be nice if that model could be checked for validity.

Materials scientists have tried to measure the strength of the bond between layers of atoms by actual experiment, essentially pulling on a brittle specimen, and measuring the force and elongation at fracture. Knowing the spacing between atoms, the force to break the "springs" could be calculated.

Based on the simple "spring" model, scientists have tried to calculate the stress that would just separate two adjacent layers of atoms, which would presumably be the stress at failure for a given material. It is interesting to note that the *theoretical* breaking *strain* for a very wide range of materials is about 10-20%, which, of course, very few practical materials actually achieve. The breaking strength, then, should be about one tenth to one fifth of the Young's modulus, and we don't achieve a test figure this high in real life.

Griffith's Model Material: Glass Fibers

So things aren't nearly as strong as they "ought" to be! Why not?

Shortly after World War I, at the Royal Aircraft Establishment at Farnborough, A. A. Griffith, a materials scientist, tried to answer this basic question. The most obvious experimental material might have been steel, but steel is actually a complex structure of different interlocking crystals and uncertain crystal boundary states. Ductile steel usually fails by having crystals *slide* past each other, rather than by breaking the fundamental atomic bonds. Griffith wanted to study a model material which had a simple *brittle* failure.

He chose common glass.

Using the simple "atom ball and spring" model, and the measured Young's Modulus, Griffith predicted that the strength of common glass should be close to 2 million psi at room temperature. Actual glass rods of about one millimeter thickness broke at 25,000 psi - quite a bit short!

He then heated the test rods and drew them down to smaller fibers, then continued the tests. At two thousandths of an inch, he tested near 50,000 psi. At one thousandth, over 100,000 psi. At half a thousandth, nearly 250,000 psi. At around a ten-thousandth, he got 500,000 psi. Extrapolating to a fiber of zero thickness gives an ultimate strength of 1,600,000 psi, close enough to the prediction, considering the rough model!

We conclude that the "atom ball and spring" model isn't too bad when we are just stressing a tiny area of not too many atoms at a time. In the real world, where we must use lots of atoms at a time, other factors make the specimen fail.

Why Things Break

Suppose we could make a perfect "large" crystal of a test solid, in which the "atom ball and spring" model applied exactly. The surface of this ideal specimen would be perfectly smooth with all the atoms lined up perfectly in layers. If we test this beautiful piece, Griffith's work suggests that we should see the true strength of the interatomic bond, even in a "large" specimen. For this idealization, we have all of the atomic bonds across the entire piece bearing their fair and equal share of the load, until they all fail at once.

Suppose we test this ideal crystal and strain it until we find just the stress that causes failure, i.e., just separates the atomic layers.

Now, let's repeat the test, but assume that our perfect specimen has a tiny scratch or flaw on the side, one atomic layer wide and several atoms deep. The atom layers on either side of the scratch are no longer attached to their neighbors. As the test stress is applied, the first *attached* bond must receive the added stress that *would* have gone to several neighboring *unattached* bonds. The bond fails. Now the next bond in line has even more stress. It fails, too. A tear then propagates across the entire specimen, started by the surface flaw. This is *brittle failure*, caused by a *stress concentration*, and happens to be the chief failure mode of glass for large practical specimens.

This is illustrated by the trick where a macho man tears a telephone book in half. The trick is to push the pages in from the side to make an arch of bent pages. Tear from the bottom and the "strong" man needs to break only one page at a time at the stress concentration!

We can extend our ideal test thought experiment to another less obvious real life case: Suppose our perfect crystal does not have a scratch extending into the body of the crystal, but, instead, has a *step*. A *step is any sharp discontinuity of thickness*, where the specimen abruptly goes from one uniform thickness to another.

If we load a specimen which has a step, the thicker section (larger area) must have less stress than the thinner section (less area), because the *load* is the same. Now see that the stress at the outer edge of the thick section *must turn at the step and add to the stress* already carried by the very first atomic layer *at* the step. Like with the scratch, this first atomic bond will fail at the *stress concentration* and the brittle tear propagates in the same way as with a crack. The specimen fails where it is assumed to be "stronger!"

Fracture Mechanisms

Materials break by one of two mechanisms, and which happens first depends on the material:

1) *Brittle cracking,* by propagation of a tear crack, as we have just discussed, and

2) *Plastic flow*, where atomic layers slide past each other and re-connect when the strain displacement ends.

We say that a material is *brittle* if it fails first by cracking. (Bulk glass and porcelain are brittle.)

We say that a material is *ductile* if it fails first by plastic flow. Gold is extremely ductile. Copper is very ductile, but less so than gold. Ductility is generally associated with metals, and is very useful because we use that property when we draw copper wire, stamp steel automobile fenders, or hydroform a curved airplane engine nacelle from aluminum.

For the particular case of glass, which is going to be very important to our discussion, the chief limitation to the useful strength of bulk glass (beer bottles, windows) is (brittle) tearing failure from surface flaws that are impossible to avoid. By going to thin glass FIBERS to reinforce a composite, we avoid the consequences of brittle failure and force the failure to occur by plastic flow at a much higher stress.

Note that permanent deformation by plastic flow is still a service failure of a part, even if the deformed part has not actually separated into pieces. (*If it is bent too far to serve its designed function, the part has failed.*)

> *We have such an easy familiarity with ductile steel in our lives, that we have a somewhat permissive attitude toward incidents of catastrophic failure in steel structures. To a degree, a failed ductile part, such as a propellor blade or landing gear strut, might be bent back to its original position and still serve safely, though this is almost never a safe procedure in aircraft service.*

Whiskers - a Curiosity

It is possible to grow tiny pure crystals of many materials in the form of long smooth filaments, "whiskers", thousands of times longer than their thickness. When tested, these tiny crystals usually turn out to be extremely strong. The surfaces of these grown crystals are generally quite smooth and free of cracks, and the interior is free of defects. Whisker crystals have been used as reinforcement in experimental composite materials to produce quite spectacular (but quite expensive) materials. The growth process does produce steps where the crystal build-up is not uniform. Whisker failure is found to occur at the stress concentration caused by these steps.

Toughness - or Resistance to Brittle Fracture

Considering our discussion this far, it seems a wonder that we can ever find a strong material. After all, we'll never be able to get perfect materials and we'll never avoid surface flaws, so why doesn't everything we own tear like Cellophane from the inevitable flaws? The answer is that practical "strong" materials are those that possess the quality of *toughness*, or *resistance to failure from propagation of cracks*. As aircraft mechanics, we drill "stop holes" at the end of obvious cracks in aluminum to limit that propagation. But there are plenty of other smaller flaws that are not obvious. To get toughness, we must stop cracks.

Work of Fracture

It is easy to ride a bicycle on a hard, dry pavement. It is very difficult to ride in deep mud. The difference is that considerably more energy must be expended to deform and fracture the mud than is required to just flex the tires a little and turn the bearings. This difference appears to the bicycle rider as resistance to her progress and to the materials scientist as increased *work of fracture*.

Nature always takes the action that leads to a lower energy state, the "easy way out". When considering the flow of energy, "if it *can* happen, it *will* happen". This rule includes the propagation of cracks. If the energy of the system is reduced by propagating a crack, then a crack is what we'll have.

Griffith observed that there are two conditions for crack propagation:

1) The energy of the system is reduced by a crack, and
2) There must be a mechanism for the release of the energy.

If these conditions are not met, there will be no crack propagation under load and we may say that the material is *tough.*

Strain Energy and Surface Energy

Let us apply Griffith's crack energy concept to a steel clock spring like those Mr. Hooke studied. Basically, a flat steel clock spring is a beam. When we wind it up, we are bending the beam the same way we do when we stand on the end of a diving board and make the board bend down.

As we apply force to a clock spring, or any other beam which follows Hooke's Law, the force required increases directly as the strain increases. Force times distance is defined as *work* and we get this work back if we let the spring unwind to drive the clock. But the *definition* of energy is the capacity for doing work: we store energy in a clock spring by deforming the little atomic springs in the volume of the material.

So there is strain energy in the wound (strained) spring and the energy per unit volume of the spring material is:

1/2 stress x strain

So why doesn't nature take the easy way out and propagate a crack from the first handy tiny flaw so that we can't have clock springs or airplanes?

The answer is that there is a competing energy state for the clock spring: *surface energy. The crack will propagate if it takes less energy to make new surface area at the crack than is available in strain energy at the crack.*

The atoms *inside* a solid are happily attached in every direction to their neighbors. A surface is different; *there are no neighbors on one side.* To form new surface, all of the broken atomic attachments of the atoms on each side of the new surface must be formed again with neighboring atoms or atoms dragged from the interior. This re-quires repositioning of surface atoms and takes energy, which is as-sociated with the new surface.

Suppose we have three tensile test specimens, all under stress, as shown below. The left hand specimen has no crack and the stress lines are evenly distributed. The middle specimen has a small crack and the right hand specimen has a large crack, so the stress lines are distributed as shown.

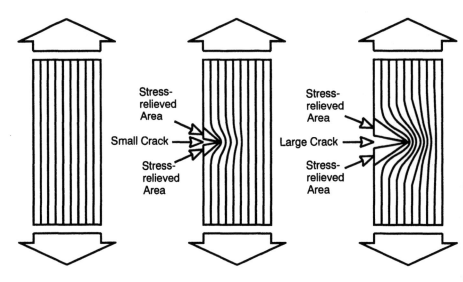

In each case, the stress lines are concentrated in the vicinity of the crack. See that the material on each side of the crack is *no longer strained* (the crack has opened, relieving the strain), so there is no corresponding stress. The strain energy formerly stored in the re-lieved volume of material is available to propagate the crack. If the work required to make new surface is less than the strain energy re-leased by the crack, Nature has her easy way out and the crack will grow to failure. This required work is the *work of fracture*, the same as illustrated by our example of riding a bicycle in the mud. If the work of fracture is too great, there will be no propagation.

Whether a crack propagates or not depends on the original size of the crack, as well as the qualities of the specimen. If the crack is very short, there is only a small volume of material which is relieved of strain, so there is only a small additional energy to contribute to the tear.

Consider the large crack on the right, in our sketch. Now the crack is twice as long, but the volume of material relieved of strain is *greater* than twice as much, approximately proportional to the *square* of the crack length. Eventually, this "square law" will catch up with us, when the strain energy will be enough to meet the work of fracture rule, and the tear will propagate. In the simple case, there will be a critical crack length, which is called the *critical Griffith length*. Below this limit, the crack is stable; above it, the specimen tears.

Actual numbers for the critical Griffith length depend on the particular material and the application. We see, though, that for toughness in a material (large work of fracture) there is some mechanism where lots of energy is absorbed *far from the fracture surface*. Our bicycle-in-the-mud analogy illustrates this; the rider pedals hard to fracture the mud, but the actual fracturing takes place many inches from the pedals, due to the mechanism of the bicycle! Wood and steel are particularly good examples of tough materials because their failure mechanisms involve movement of cell walls or crystal boundaries over a large volume, far from where the load is applied. Give Nature too much work to do and she won't do it!

Strength and toughness are inherently mutually exclusive. Much of composite design is an attempt to find solutions to this compromise problem.

> *Toughness is not the same as strength. We need both, and the art of material selection requires inevitable compromises.*

Chapter 2
Forces In Structures

Very Elementary Beam Theory

A beam is a structure that transfers loads between locations. If we stand on a wooden plank to cross a ditch, the plank is a beam that transfers our weight to the banks and keeps us out of the ditch. A more complex highway bridge across a river does the same thing. An aircraft wing is a cleverly designed beam that receives the loads from the fuselage and transfers those loads to the air (or the other way around - it doesn't matter). The simple landing gear struts used on the smaller Cessnas are beams. Beams make up a large proportion of the devices of ordinary life - and certainly of aircraft structures - and it is useful to have an intuitive idea of how beams work. If we have a basic understanding of elementary beam theory, we can extend the that understanding to the more complex structures used in composite aircraft.

A Simple Case: The Cantilever Beam Panel

We will discuss the example of a *single load* at the end of a beam where the beam is firmly attached to a wall. Engineers call this a *cantilever* beam.

Let us consider two very simple cases of suspending a weight W, of 100 pounds, from a structure.

First, consider the case of a weight W1 which is supported from a point P1, which is attached to a wall at A1 and B1 by the structure shown below.

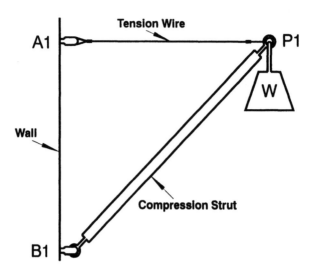

The Cantilever Beam Panel - Case 1

Our structure is ideal because the strut and the wire are weightless and the pivots have no friction. The wire can only act in tension and the strut can only act in compression.

Consider all the forces acting on point P1, where the weight is sus-pended. Point P1 is stationary, so all the forces must be in balance. Weight W exerts a *downward* force, so we need an equal *upward reaction* force to balance it.

But we don't have such an upward reaction force directly - we only have the combination of the strut reaction upward at a 45 degree angle and the tension wire reacting horizontally. The sketch on the next page shows these forces.

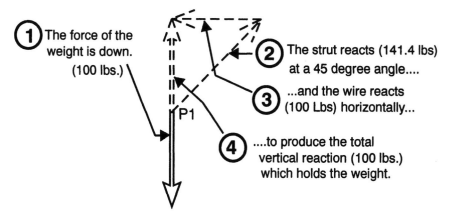

① The force of the weight is down. (100 lbs.)

② The strut reacts (141.4 lbs) at a 45 degree angle....

③ ...and the wire reacts (100 Lbs) horizontally...

④ ....to produce the total vertical reaction (100 lbs.) which holds the weight.

The Balance of Forces Acting On the Weight at P1

A force is a *vector*, which is a quantity which has both magnitude and direction. *(Temperature, in contrast, has only a magnitude, so it can't be a vector quantity.)* A force, then, can be represented by an arrow with a length representing the magnitude (100 pounds) and a direction (down, for our weight). For the example in the sketch, we see that the upward reaction force (4) (100 lbs.) that holds the weight from falling is the sum of the reaction of the strut (2) (141.4 lbs.) and the reaction of the wire (3) (100 lbs.). The vector sum (4), (100 lbs., up) is found by placing the head of one vector at the tail of the next to find the resulting vector, as shown in the sketch.

For this brief discussion, let's use the convention that a *driving force that tends to cause motion* in its direction will be represented by a *solid-line hollow* arrow as in the example above at (1).

If a force is caused by the structure to have *components* in other directions, say to compress the strut or to put a tension load on the wire, these components will be represented by *solid-line plain* arrows.

So solid-line arrows represent the *driving* forces and their components which would cause the structure to move (strain) in the direction of the force. Ultimately, this is the force that could break the structure.

Our ideal structure deflects a tiny bit (strain) and generates a reaction force (stress) which opposes the force of the weight and successfully supports it. We'll use the convention of *dashed* arrows to show *reaction* forces.

We see by scaling the arrows that the *horizontal* tension load on the wire is 100 pounds, the same as the *vertical* load from the weight, and the compression load on the strut must be 141.4 pounds. The forces are in the directions shown.

Now let's look at attachment point A1. The force applied by the wire which would tend to pull out the attachment at A1 and the re-action to that force is shown here:

Similarly for attachment point B1: The force transmitted by the

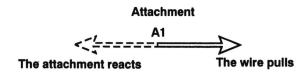

Attachment

A1

The attachment reacts   The wire pulls

The Forces on Attachment A1

strut is shown below as a solid arrow, with components down and to the left, countered by the opposite reaction forces from the struc-ture (dashed arrows).

So if we draw all the driving forces (solid arrows) and reaction

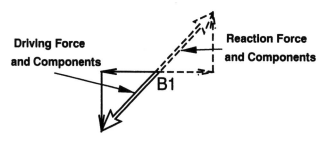

Driving Force   Reaction Force
and Components   and Components

B1

The Forces on Attachment B1

forces (dashed arrows) seen at points P1, A1, and B1 for the first case, we get the diagram on the following page:

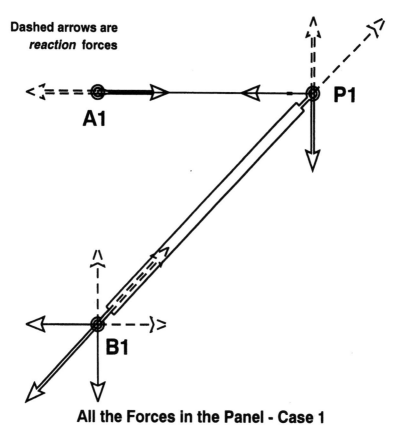

Dashed arrows are *reaction* forces

All the Forces in the Panel - Case 1

Now that we see how reaction forces work, we'll leave them off of the diagrams to follow. If a driving force is shown and the structure isn't moving, we will know that there must be an equal reaction from the structure.

The Simple Beam Panel - Second Case

Now let us introduce a second similar structure to support a second 100 pound weight, W2, sketched on the next page, with the roles of the strut and the wire reversed from the first case:

Clearly, this is just as good a way to support the weight as the first case. Let's support the 100 pound weight from point P2, and let's let the corresponding attachment points be A2 and B2. By an argument similar to that we used for the first case, we can see that the forces at P2 and at the attachment points A2 and B2 will be as shown.

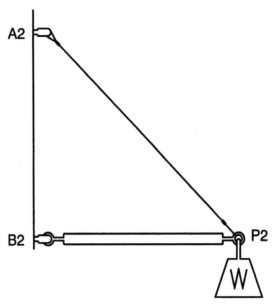

The Beam Panel - Case 2

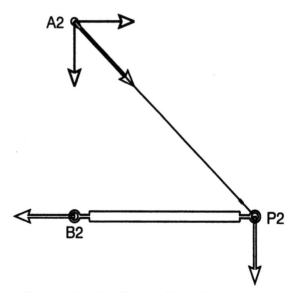

Forces in the Beam Panel - Case 2

The Simple Panel - Assembled

Now let's take the first structure and the second structure and connect them both to the same attachment points, which we'll now just
call A and B. The two structures together make one *panel* of our
ideal beam. *The forces from our two structures add at A and B and*
we get the combination shown below:

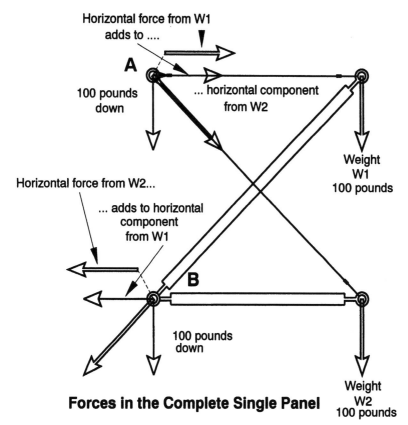

Horizontal force from W1
adds to ....

A

100 pounds
down

... horizontal component
from W2

Horizontal force from W2...

... adds to horizontal
component
from W1

B

100 pounds
down

Weight
W1
100 pounds

Weight
W2
100 pounds

Forces in the Complete Single Panel

Notice that the horizontal members - whether the wires in tension
on the top or the struts in compression on the bottom - experience
very real forces, *but do not directly lift the weight.* The vertical forces
that must support the weight are provided only by reactions in the
diagonal members caused by strain in the material. *All of the mem-*
bers are essential; if any one fails, the weight must fall.

See also that the *horizontal* forces at both attachments are equal to twice the force of one weight, but the vertical force remains equal to the force of one weight.

Extending the Panel to Make a Beam of Several Panels

But the points P1 and P2, where the weights are actually attached, are just like points A and B which receive loads through the structure. Let's add a new identical panel to extend our beam, so that the new panel section is attached to the base wall at C and D and receives *its* load from A and B. (*Engineers call this structure a lattice girder.*) The new panel experiences strain in reaction to the loads from the previous panel and our loaded beam now looks like this:

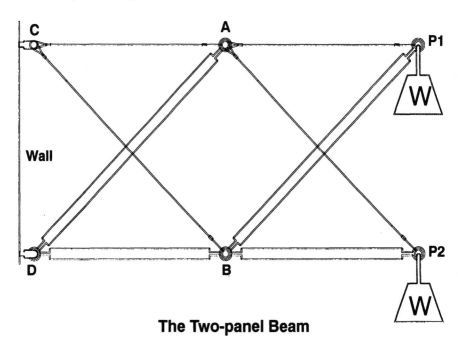

The Two-panel Beam

And if we use the same argument we used for the single panel case, the new vector diagram for the beam looks like the figure on the next page.

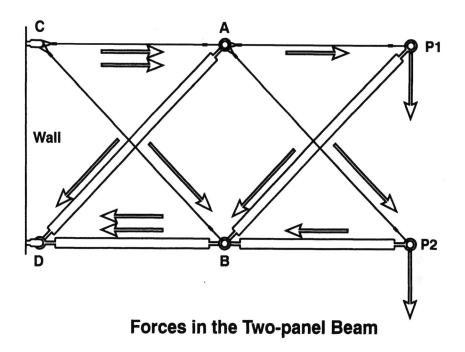

Forces in the Two-panel Beam

(We're cheating a little here by showing as many "unit" arrows in parallel as there are "unit" forces in the member. To be correct, we should show the vector addition as arrows head to tail, but then they wouldn't fit in the diagram.)

Notice the result of doubling the beam length. The forces at the attachment points due to the tension and compression in the horizontal members have doubled, but the forces in all the diagonal members are the same. *See that the diagonal members of each panel feed load to the horizontal members and that adding the new panel adds a unit of horizontal load, but does NOT increase the load in the diagonal members.*

Finally, below we have the case for *four* panels in our simple example beam. The forces then are as shown below:

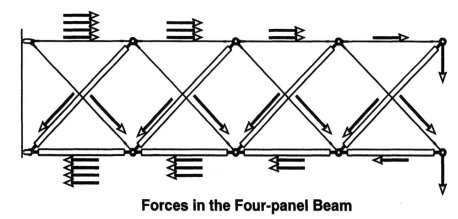

Forces in the Four-panel Beam

And we can draw some conclusions:

1) As we extend the simple cantilever beam, the tension and compression forces in the top and bottom of the beam increase in proportion to the length of the beam.

2) The forces in the diagonal members of the cantilever beam structure are *constant*, independent of the length of the beam.

To apply our reasoning thus far to a practical beam design for this simple case, we see that we will have to provide more material in each new horizontal section of each panel to carry the increasing loads at the same stress, *but we will not need to add material for the diagonal sections.* If we were to make this beam from a composite material, we might use an additional layer of glass for tension and compression in each new panel, so that there would be as many layers at the attachment end of the beam as there were panels, but we need to use only *one* layer for all the diagonal sections. We would then have the *same stress* throughout the structure.

> *Finally, remember that we have been discussing just an ideal, flat, two-dimensional structure that can be pictured on a flat piece of paper. A real beam must have a finite width or some means to oppose forces in the third direction. We'll discuss this in Chapter 4.*

Multiple Loads

We can extend our simple analysis to include the general case of many loads placed along our lattice girder, instead of just being applied at the end. We know how to calculate the forces in a single panel of our lattice: All we have to do for the general case is to place weights along the lattice at the panel junctions as we please and add all the resulting forces (as vectors) to find the result. We can analyze the case of *equal* loads all along our lattice girder example by placing equal weights at all the lattice connections, calculating the contribution of each weight, and adding all the forces. Engineering handbooks show the forces in simple beams for a number of ideal cases.

Most beam loads are *not* ideal, however. Cars on a bridge don't space equally and many loads change continually along the span, as with an airplane wing. For a wing, the distribution of lift (usually) goes from a maximum near the root to zero at the wing tips. (And the landing gear will probably put an impact load at some discrete point on the wing.) To apply this load distribution to our lattice beam, the load can be approximated by dividing the total lift into a group of equivalent separate forces which are applied to our lattice structure at the junctions.

Bending

We calculated the forces at the attachments of our simple lattice beam by starting with the applied load and following the forces in each member. We saw that our beam, which was four times as long as it was wide, developed horizontal forces at the attachments which were four times the applied load. This member-by-member form of analysis is rarely practical or even possible in actual engineering design.

Suppose that we recognize that the most highly stressed part of the beam is at the attachment and that we do not care to calculate the horizontal forces elsewhere in the structure; we only want to know the forces at the critical point, the attachment.

The sketch below shows our same beam, but without calculating the horizontal forces in all the members. The loaded beam acts as a lever on the top attachment, similar to the action of a carpenter's claw hammer in pulling out a nail. The bottom attachment is similarly pushed in by the pressure on the line of horizontal struts.

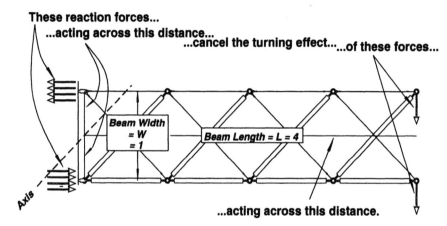

Turning Moments at the Beam Attachment

The combination of horizontal forces acts to *bend* the beam at the attachment and the bending influence is the product of the applied force and the distance from the axis. Structural engineers refer to this bending influence as the *moment* on the beam. We see that our two unit loads are applied to our beam which is 4 units long and one unit wide.

We see that our two unit loads are applied to the right side of our beam. The beam is stationary, so the bending moment is opposed by the four unit reaction forces that we know are developed at each attachment. The beam is symmetrical, so the axis of the turning forces is in the middle, as shown. The turning influence at the attachment is then a force of magnitude four pulling above the axis, acting over half the beam width, plus a force of four pushing below the axis. That's eight unit forces acting over beam width 1, opposed by two unit forces acting over beam length 4. We have calculated the horizontal forces at the attachment without calculating the forces in the intermediate members. In this case, the *bending moment* is four.

Structural engineers make extensive use of bending moments in designing complex practical structures.

Shear

Carrying Loads Across a Beam

So far, we have made a simple beam from ideal members which permits us to analyze the forces exactly. We showed that we can hang a weight on the end of the beam and trace the forces all the way back to the root of the beam. We see that the network of crossed *diagonal* members - alternately in tension and compression - actually provides the *vertical* force which holds up the load and transfers forces to the horizontal members. For this example, we chose to use diagonal members at an angle of 45 degrees, but some other angle - say 30 or 60 degrees - could have been used just as well. We could have drawn our arrow (vector) diagrams for this case and calculated the forces involved. The essential idea is that if the beam is to function, *real forces must be transmitted between the top and bottom horizontal members*. We provided ideal diagonal members to do exactly that. Engineers call the structure we have been discussing a *determinate* structure because all the forces can be calculated.

But real life is not that tidy. We see that real airplane wings rarely use a lattice girder like ours for a wing spar. A spar might be a solid piece of material, such as a piece of spruce wood with no obvious "horizontal" or "diagonal" members. Our network of diagonal members might be replaced by a solid aluminum sheet or by a sheet of aluminum with circular lightening holes cut out. (Clearly, the air in the cut-outs cannot carry any load!) In composite aircraft, our 45-degree lattice might be replaced by continuous layers of glass cloth set so that the yarns in the weave lie at approximately 45 degrees to the spar, like in our example, but there are now hundreds of yarns. These are *indeterminate* structures. The point is that the forces between the the compression members and the tension members may be transferred in a complex way in a loaded beam, but they *are* transferred.

> *The structural engineer might say that the difference between a stiff steel rod and a flexible steel stranded cable the same size is that the separate strands in the cable can't pass shear forces, so the cable "beam" bends.*

Shear Force In Practical Beams

If we replace the diagonal members in our simple lattice girder structure with a continuous structure, say an aluminum sheet, the load is still supported, but it is no longer obvious how the forces in the beam are distributed. We could actually build the physical structure, place a few hundred strain gauges on it, load it, and observe where the internal strains (and the corresponding stresses) occur. We would then see the actual directions and distribution of forces.

The sketch below shows our ideal lattice replaced by a solid sheet of the same size. The beam is still four units long and one unit thick. Here we'll go along with many engineering texts that discuss beams

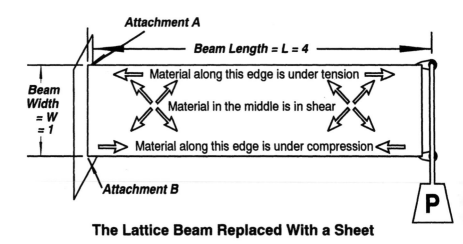

The Lattice Beam Replaced With a Sheet

and we'll use "P" instead of "W" for the applied force and "W" to show the width of the beam.

Engineering handbooks show the vertical load transferred from one side of a loaded beam to the other as if it were expressed as the cutting (shearing) action of a scissors and they express this as *vertical shear*. Average vertical shear stress, then, is the total vertical load at a point on the beam divided by the cross section area at that point. This is often not very meaningful, because we need to know where the *maximum* stress is, not the average. In practice, we overcome this by using large safety factors and using the *average* as a rough guide to determine the *maximum*.

There is a theorem in mechanical engineering which shows that wherever a body is subject to shearing stress along one plane, there must be an equal shearing stress in the perpendicular plane. In our simple lattice girder model, the diagonal tension wires and diagonal compression struts *are* perpendicular to each other and the principle is shown. A very good way to think of shear stress is as a continuous series of tension and compression forces at 45 degrees to the axis, like a lattice.

A Side Issue: Shear as Twisting a Basket

Historical note on the concept of shear: There is considerable evidence that the builders of wooden sailing ships went on for some centuries without actually grasping the concept of shearing forces and the consequences of these forces on their designs. Frames and planks were essentially placed at right angles, so the structure was rather like a basket. This gave a flexible structure that could accommodate the swelling of wood in water and - like a basket - distribute the loads placed upon the hull and rigging. Modern engineers see a ship as a shell structure subject to (plenty of) bending and torsion. The traditional wooden ship was built like a wooden garden gate which had vertical and horizontal planks but didn't have a diagonal member. The result was that the shear stress was taken by the caulking of the seams which alternately closed and relaxed under load, like wringing out a wet towel, inevitably causing wear and leaks.

Sir Robert Seppings, of the Royal Navy (1764-1840), finally clearly understood the problem about 1830 and introduced diagonal iron bracing into wooden naval hulls. In Victorian times, these "composite" hulls (yes - their name!) became common in the navy, but most wooden ships continued to be built in the traditional way, without the bracing. Some designers are still slow to get the word.

Buckling of Columns

Airplane designers from the Wright brothers through the wood and wire days of World War I and down to modern times have tried to use materials in tension wherever possible. This preference is explained by the fact that you can pull a substantial load with a slender light weight wire, but you certainly can't push one. Eventually, all the load vectors must be gathered and delivered to some other part of the structure that can stand compression without failing. The problem is that structures that must be compressed usually bend (buckle) before they fail by crushing, and never achieve their theoretical compression strength.

A civil engineer can take a short, stubby sample cylinder of concrete, place it in a testing machine, and compress that sample to destruction. The engineer can then draw a conclusion about the tested compressive strength of that sample and can be guided by the result in using the concrete in a practical design.

If the compression test sample is *not* short and stubby, but is long and thin like a practical airplane strut would be, it nearly always bends into an arc before failing. The average stress at bending failure is very much less than the stress that would be measured with a short specimen of the same material. Engineers call this bending failure mode *buckling of a column*.

A Side Issue: Buckling Demonstrated

A practical illustration of buckling: It happens that the ultimate strength as a column of an empty common 12 ounce aluminum beverage can is just about equal to the weight of the person who drained that can. If that person wishes to give a dramatic illustration of the phenomenon of buckling, he/she may carefully stand on the end of the empty can and announce the experiment to nearby friends. If sufficiently agile, the experimenter may lean over very carefully and gently tap the walls of the can with the finger tips. The can will collapse very quickly as the structure becomes unstable, so it is important to get the fingers out of the way.

Buckling Determined By Shape

The practical design of long compression columns in aircraft construction is best left to specialists, though it is certainly practical for an amateur to make up a test column and load it to destruction. In fact, for really complex indeterminate structures loaded in compression, amateurs don't do much worse than the gifted engineers at Boeing. Even these must go to the trouble of placing many strain gauges in a prototype structure, applying test loads, and observing the resulting stresses. Whether the designer is an amateur or a professional, both must test a complex design and be the wiser for the answer given by Reality. Modifications will usually result from an actual test.

The simple column design formulas given in engineering handbooks give useful guidance to show what is important. For the case where the column is a thin hollow tube (as is a beverage can), the allowable compression load is directly proportional to the modulus of elasticity (E) and proportional to the *fourth* power of the radius of the tube, and diminishes with the *square* of the column length. Clearly, high material stiffness (E), is good and unsupported length is bad, but adequate column radius compensates very powerfully (4th power!) for the length of the column. This is why aluminum sailboat masts are hollow tubes and why sailboat rigging is designed the way it is. Horizontal spreaders are placed at an intermediate level on the mast to convert the mast to the equivalent of two shorter stable columns end to end, rather than one very long unsupported column which would buckle. Only the very smallest sailboats can use a mast without spreaders.

A key idea is that buckling starts at a part of the structure which has a displacement at a right angle to the compression force. This displacement (called eccentricity) doesn't need to be very large to start failure. This is illustrated by the experiment mentioned above with the aluminum beverage can: Sometimes, if the can has a small dent or other flaw, it collapses before the experimenter can successfully stand on it, forcing the experimenter to drain another test can and repeat the experiment.

Chapter 3
The Nature of Composite Materials

An Introductory Puzzle

The following simple puzzle illustrates the effect of using two materials of *different stiffness* in the same structure. The solution leads directly to an understanding of why composite materials work as they do.

Consider the problem sketched below. We have a 1,000 pound weight, a steel chain which breaks at 600 pounds, and a rubber bungee cord which also breaks at 600 pounds. The problem is to suspend the weight as shown, using both the chain and the rubber cord. (No tricks, like doubling the chain, are permitted or necessary!) The weight may be lifted into position by any means and any desired measurements are permitted. Recognize that simply paralleling the chain and the bungee won't work: When the weight is released, nearly all of the load first goes to the stiff chain and breaks it. All load then goes to the flexible rubber, which now cannot carry it, and down comes the weight. How must it be done?

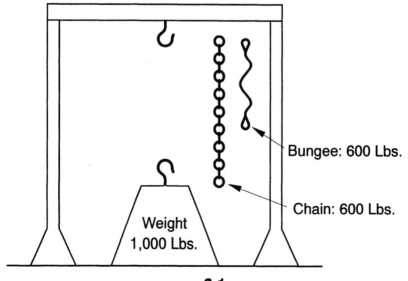

Bungee: 600 Lbs.

Chain: 600 Lbs.

Weight
1,000 Lbs.

The answer is this:

1. Connect the rubber bungee and lower the weight until the rubber carries a load of, say, 550 pounds, known to be a safe level of stress.

2. Connect the steel chain and continue to lower the weight until both rubber and chain are carrying the full load of 1,000 pounds. The bungee now carries about 551 pounds and the chain carries the remainder, 449 pounds.

See that the rubber bungee must be given a large displacement (strain) if it is to carry its share of the load. The chain, being stiff, requires only a very small additional strain to carry its load. The bungee is only slightly affected by the additional strain required to load the chain.

This illustrates a principle that carries into practical composite aircraft design. Some designs which use mostly E-glass fiber make use of carbon (graphite) fiber for spar caps. Carbon is much stiffer than glass, so the carbon will carry nearly the entire spar cap load. If any glass is used in parallel with the stiffer carbon, the glass component is essentially wasted because it will never experience enough strain to carry a useful load - the carbon fiber gets nearly all the load, just like in the puzzle with the stiff chain and the soft bungee.

Two-phase Materials

In the early 1960s, workers at the Owens-Corning Fiberglas Corporation took a very general view of the basic principle behind composite materials. They described any combination of a strong and stiff material embedded in another less stiff material as being a *two-phase* material. There are many examples in nature and in engineering. (Two-phase materials are very old in metallurgy.)

The requirement for a practical two-phase composite material is that the "strong" component must have both a high breaking strength and a relatively high modulus of elasticity. The "matrix" component must have a lower modulus of elasticity (generally in the range of one quarter that of the stiffer material) so that it can deform to distribute stress to the "strong" component.

Bamboo - a special case of wood - is a particularly good example. Ordinary dry bamboo consists of densely packed cellulose fibers (Young's modulus, E, about 4 million psi) in a of a plastic-like filler of lignin (E = 1 million psi), with distributed air spaces. Older fishermen will agree that bamboo made excellent fly rods for many years. Modern fly rods, at any price, are now nearly always made of a composite of carbon fiber in a plastic matrix, a combination which outperforms bamboo, giving greater stiffness for the weight. It would be theoretically possible to make a fly rod out of steel, but the author knows of no serious attempt to do this.

Another example of a two-phase material is tough metal alloys where crack propagation is stopped by crystals of different material embedded in the mass of the specimen. Even a high speed grinding wheel is a good example: the alumina abrasive (E = 50 million psi) is held in a ceramic carrier (E = 6 million psi) and the resulting structure does not crack as easily as either component alone.

Two-Phase Material Combinations

When we discuss practical two phase composite materials, we usually imply that the "strong" component is present in units (crystals or fibers) that are quite small compared to the size of the structure. In the case of glass fibers, the small "units" can be considered to be the undamaged sections between the inevitable surface cracks in the fiber.

Some combinations do not work because the elasticities do not have the right ratio. A mixture of glass fibers in portland cement does not work because the stiffness of the two components is nearly the same. If we use steel in the same portland cement (familiar as reinforced concrete) the combination *does* work. Asbestos is a stiffer fiber than glass and asbestos fiber in portland cement was used successfully for many years as furnace flues, until the toxicity of asbestos was understood. Glass fibers in plaster of Paris works very well and is a handy material for making molds. Even water (ice) can be used as a matrix, if the builder is willing to accept some obvious temperature limitations!

Glass fibers and common tan-colored water-catalyzed urea-formaldehyde glue, as found at hardware stores, makes a very handy and useful two-phase combination. The author uses it in non-aircraft household woodworking for such things as reinforcing the corners of plywood drawers. It is cheap, sets fast, allows the use of thinner wood, grinds out smoothly, and permits a water clean up.

The geometry of the "strong" component is important. For example, glass fibers in epoxy resins work very well as composites and we will have much to say about this combination. The same glass as a crushed powder in the same resin does not work, as will be made clear below. Also, the tiny hollow glass microspheres (generally called "micro"), when mixed with resins, are widely used as an excellent light weight filler in aircraft construction, but are NOT suitable for load-carrying applications.

How Two-Phase Composites Work

To clearly understand the basic mechanism of a practical two phase material, first consider an ideally perfect structure of absolutely flawless fibers, maybe a rope made of Griffith's unscratched fine glass fibers which can reach enormous strength. Further, assume that we can make these flawless fibers as long as we please. Let us make a full size test specimen out of this perfect glass rope and strain it in a test machine. We find that each fiber carries its full theoretical share of stress from end to end and we conclude that we have one strong specimen indeed!

Now repeat the test with a rope section made of real-life glass fibers, the kind that always have many surface defects that propagate cracks. We find that we are back where we were, discovering how easy it is to propagate cracks in glass where the surface cracks exceed the critical Griffith length. The real glass rope is useless.

Let us next use a real-life "strong" material - glass, again - but this time embed the specimen in a matrix material which has a lower

Young's modulus than glass, like a plastic. If the glass is perfect, we would have the first case sketched below. The stiff fibers and the less-stiff matrix both have the same strain and nearly all the load goes to the glass. The sketch shows a series of reference lines which remain parallel because all parts of the specimen - glass and matrix - have the same strain.

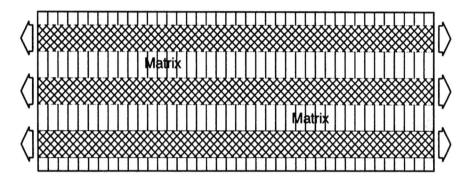

Next, consider the case sketched below where the middle fiber has a flaw that allows the fiber to break into two sections and where the intact sections are long compared to the diameter of the fiber. Also, we assume that the two outer fibers remain intact and that there are no other flaws exposed nearby.

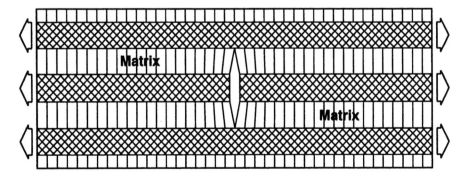

See from the sketch that the broken fiber has *no strain at the break*, so the *stress* is zero, but the attachment between the matrix and the broken fiber remains good. As the glass fiber is relieved of strain at the break, the broken ends move slightly and a *shear* stress is placed on the attached matrix. This shear stress is at a maximum at the break and approaches zero at a distance from the break.

Recognize what happens here: The two-phase material exchanges the shear stress of a much larger area of matrix material for the tensile stress of a small area of fiber material. The trade is a good one if the "strong" fiber doesn't have too many breaks that have to be covered by the shear loaded matrix. This is why we can't use ground glass fragments in the matrix, but must use proper, mostly-intact long fibers.

The shear stress in the matrix acts to transfer the tensile stress - which would have been carried by the broken fiber - to the neighboring fibers.

The sketch below shows a closer view of the region around the end of a fiber section and shows a second mechanism for passing stress around a flaw in a fiber. When the fiber breaks, and forces a crack in the matrix, the crack may propagate only a limited distance and stops in the matrix by itself if the conditions, such as the work of fracture of the matrix material, are favorable. In this case, the matrix is inherently crack-stopping and it is not required that the crack stop at a nearby fiber.

Crack stopped at adjacent fiber

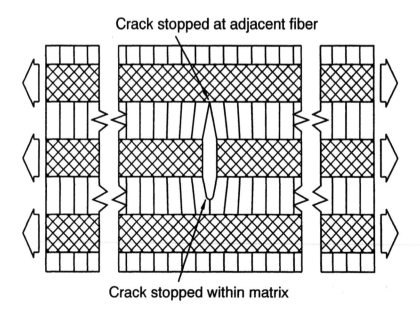

Crack stopped within matrix

In this idealized two-dimensional sketch, the strain in the matrix around the break in the fiber is shown as being resolved in roughly two diameters of the fiber. Whether this is closely true in a real life case depends on the nature of the structure. An actual three-dimensional case is much more complex and the spacing between adjacent fibers will vary, but the principle is illustrated.

Crack Stopping and Toughness

It is crack-stopping that gives us toughness. The composite structure has much of the strength of the normally brittle fiber, but gains the toughness of the matrix. The work of fracture of the structure can be very large because, to break the part it would be necessary to break the fibers or pull the fibers out of a large volume of the matrix. For tension loads, it is not even necessary that the adhesion between the fiber and the matrix be as strong as the matrix. Crack-stopping in tension can be aided by local separation between the fiber and the matrix. The crack goes to a fiber and is stopped by propagating the crack along the fiber surface when the bond between the matrix and the fiber breaks.

The tail assembly of the Boeing 777 uses carbon fiber as the principal material. The matrix is an epoxy which is "toughened" by including what is essentially finely ground rubber. We can see now that the low modulus rubber in the intermediate modulus matrix makes what amounts to tiny round voids. These "voids" work like the "stop holes" which are drilled in aluminum panels or plastic windshields to limit sharp cracks. The result is to make the final structure extremely tough, even though carbon fiber by itself is relatively brittle.

Getting toughness in a composite structure by adding material which has no strength of its own is very old technology. Ancient civilizations added straw to sun-dried bricks to stop cracks and low-tech societies still do. Cheap Bakelite thermoplastic was once very widely used to make small electrical components. By itself, pure Bakelite is too brittle for easy handling, but if very fine sawdust (cellulose grains) is added for crack-stopping, the product becomes even cheaper, but then it has enough toughness to be practical.

Compression Loads

So far, we have mostly discussed the case of loading a composite test specimen in tension. For a compression load parallel to the fibers, the specimen will fail by buckling and shear. Ideally, the fibers should be as stiff as possible to resist buckling, and stabilized against bending as much as possible by the surrounding matrix. Compression strength, then, requires that the bond between the fibers and the matrix be at maximum to realize the maximum shear stress.

In practice, we find that most failures of aircraft structures start with some component which has failed in *compression,* not tension. Compression loads are usually the design limitation and will be seen to require more material than corresponding tension loads.

Ductility and Yielding - Failure of a Structure

The sketches in the discussion above showed the cases where the composite test specimen is *elastic.* This means that the specimen will return to the original (unstressed) position when the load is removed and the experiment can be repeated many times with the same result. Even though a practical composite specimen has many cracks in the fibers, the bond to the matrix is unbroken and the stressed structure will return to the original position. Failure of a composite structure occurs when the stress becomes so great that fibers break and/or the bond to the marix fails, the failed region increases in size, and the specimen separates.

Metals are also elastic, but metals have the additional important characteristic of *ductility,* or plastic flow when stress increases beyond some limit. We are so familiar with elasticity and ductility in metals that we take it for granted. We know that an ordinary soft steel paper clip is "springy" when bent moderately, but stays bent if we go beyond a certain point, the *yield point.* With metals, the mass of the material consists of interlocking crystals and, at the yield stress, the crystal planes slide relative to each other and reconnect when the strain increase is stopped. The region where the *strain* increases considerably but the *stress* does not is called the region of *plastic flow* and is sketched below for a low strength steel and a typical non-metal composite. The phenomenon of plastic flow is almost unique to metals.

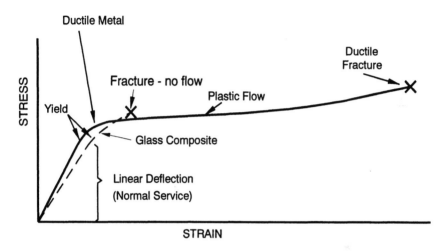

We make use of this special ductile property of metals all the time to stamp steel automobile fenders, draw copper wire, or make many formed aluminum aircraft parts. In fact, the property of ductility is one very important reason why metals are such popular engineering materials. The designer of the part presumes that the normal loads to be carried by the part will not reach the stresses which will cause yielding and, if such stresses *are* reached, that the part has failed due to overloading.

"Soft" Failure - Metals and Composites

With metals, some "failures" can be recovered from without penalty. An example of this is that it is safe and practical to pound out dents (areas which have been stressed past yield) in a steel automobile body and return the part to service. The expectation here is that no stress will come in the future which will break the part in a manner dangerous to the user. This may not be the case in aircraft service. A practical example would be the spring steel landing gear used on small Cessnas. If the gear had been bent past the yield point, it would be most unwise to just bend it straight, have a good laugh, and fly the plane again without further investigation.

The ductility of metals gives the important benefit of "soft" or partial failure, which is important for automobiles and airplanes; a damaged part may continue to function sufficiently to get the vehicle to a safe situation. We don't want either an automobile or an airplane to shatter like a teacup - we want crash-worthiness. With metal structures, there can be a very large work of fracture caused by the plastic deformation of a very large area near the failure so that parts do not actually separate and may still be partially useful.

> *Many shot-up airplanes in World War II got home, thanks to the ductility of metals which yielded near a damage site and moved stresses to undamaged areas.*

It is important to note that we may manufacture ductile metal aircraft structures by using tools to stress materials beyond the yield point, but we do not use such stresses for design loads. Composite materials do not have the property of ductility, so aircraft parts for corresponding applications using composite materials must be manufactured by other means.

Composite aircraft structures have a somewhat different failure characteristic. With composite parts at failure, there is no significant plastic flow at the yield point, but there is also a large work of fracture - the part just crushes and tears over a large volume in a messy high-energy-absorbing way, rather like wood. A failure must progress by shearing a large volume of matrix material or by pulling out many fibers from the matrix. The failure mechanism is not the same as with metal structures, but the result is much the same and may produce even greater crash-worthiness than in a corresponding metal structure. There are many anecdotes in the home-built aircraft fraternity about composite aircraft involved in quite nasty crashes which have successfully protected their crews because of the inherent toughness and soft failure characteristic of composite technology. (Very well mounted shoulder straps are necessary if the benefit of a tough structure is to be delivered to the crew!)

We have discussed the problem of designing a practical structure to safely carry some ultimate load, but few pilots routinely pull 6 gs on every turn or plonk the plane down at landing right at the maximum design load. Such high design loads may never occur in the service life of the aircraft - we just want to be sure that the design will carry them if necessary.

Fatigue Failure

When a structure fails which has been in service for a long time, but has never been loaded to near its expected ultimate design strength, we say that there has been *fatigue failure*. To be vulnerable to fatigue failure, a structure must experience repeated loading and unloading or repeated reversal of load. Fatigue is explained by the accumulation of small damage sites inside the structure with each load cycle. Eventually, after some number of repeated load cycles, the damage sites will combine to produce a stress concentration which propagates a crack and the part will fail under "normal" load. The occurrence of damage sites depends on the nature of the material and the stress. If the stress is sufficiently small, the structure may last essentially "forever." A steel hairspring in a mechanical watch may run through millions of cycles without measurable change, but working helicopter blades must be changed for safety after some specified number of hours.

Normal flight loads rarely produce the number of load and unload cycles that seriously threaten the structures of small General Aviation aircraft. A source of high amplitude cyclic loading that can threaten small aircraft is any vibration caused by a continuous resonance at some natural frequency of the structure. This is particularly true of propellers and explains why the FAA is so particular about requiring extra flight testing time if a propellor and engine combination is not certified. Long time readers of the homebuilt aviation press will recall the heroic efforts of Molt Taylor to successfully overcome the problems of complex resonances in the long propellor drive shafts required by his designs. In conventional installations, cases are reported where a propellor combination runs "forever" on one engine, but the same propellor induces early crankshaft failure on another.

Fatigue in Metals

We will briefly review the effects of fatigue in metals, because many people have had direct experience with metal fatigue. We will then extend the discussion to fatigue in composite structures.

With long-lived commercial airliners, records are kept of the number of landings and pressurization cycles that have occurred so that the level of fatigue may be estimated. The assumption is that all loads have been far below the yield point (in other words, nothing has been actually bent or plastic flow has not occurred). With metals, tiny localized changes may accumulate in the structure that could lead to failure under normal load. Fatigue occurs when the load is cyclical: applied and then removed or reversed many, many times in the service of the structure.

Fatigue in metals occurs when some single crack starts and propagates intermittently under stress, finally producing catastrophic failure. Small, localized, incremental damage accumulates as the crack propagates along crystal boundaries. Upon examination of the failed part, it will usually be found that the properties of the material in an area of normal stress near the actual failure location have been changed very little, if at all. The vulnerability of metals to cracks started by corrosion is often a factor in fatigue failure.

We hear too often from mechanics and other people who should know better that some metal part like a crankshaft has failed because it "crystallized." Being "crystallized" is the normal state of a metal. When the mechanic shows us the separate broken pieces, we see the exposed interface where the crack surface ran along crystal boundaries. The presence of stress concentration due to corrosion (which starts cracks) can be a serious contributor to a failure which is later attributed to "crystallization."

Fatigue Testing

Materials scientists who seek to understand the complex subject of fatigue must do the best they can with all these dependencies, so this is what they often do:

Groups of identical test specimens are made up and placed in a testing machine which stresses them at various levels which approximates a worst case load situation. The test load is then cycled thousands of times. The test is stopped when failure occurs. Some surviving sample specimens may be stressed to failure in a testing machine to determine their strength. The test then continues with the remaining samples, at other stresses until a stress versus cycles to failure plot is defined . The goal is to find the number of load cycles where the test specimens fail (break) at the test stress and be guided in some useful way by the result.

And there's the rub! "Design guidance in some useful way" is the problem. We are able to make only the most general conclusions. The measured fatigue life turns out to be sharply dependent on the level of test stress and that figure may relate only slightly to normal stresses encountered by a practical design. Choosing some other test stress for a reference may lead to an entirely different design.

The following effect may not relate to the problems faced by a practical designer, but is an interesting result of fatigue testing: Metals and some exotic composites can show the curious effect of "wear-in" during a fatigue test. In "wear-in", we observe a slight INCREASE in strength in specimens that have been stressed in a cyclic load. In metals, at low stress levels, the material will work harden slightly and show a greater test strength. "Wear-out" (cyclical softening) is a reduction in measured strength during the course of the test.

Fatigue in Composite Structures

The engineering literature on fatigue failure of composite materials is quite frustrating if the reader is trying to relate idealized laboratory experiments to practical designs. This problem is difficult enough with metals, but is even more difficult with composites. In order to have fatigue at all, in any material - metal, wood, or composite - there must be some accumulation of damage from repeated loading. If there is *no* incremental damage, the service fatigue life would be forever, like electrons that whirl forever in an atom.

Airframe structures made with composites will normally operate at low stresses and with very limited load cycling. Under these conditions, it is very difficult to identify a process which would start incremental damage leading to fatigue. With composites, normal corrosion is not a factor, so stress concentrations will not start from this source. With the composites likely to be used by aircraft builders, damage could, in theory, accumulate at small flaws throughout the volume of the specimen, which may reduce the stiffness slightly, but not the strength. Once the regions of the flaws are relieved, further change may drop to nearly zero and it is very difficult to predict a useful fatigue life. This phenomenon can be thought of as a "break-in" phase which has no effect on useful service. However, the repeated stress necessary to see this effect would have to be much higher than the normal design loads.

Wood is a composite. Musicians encounter fatigue in wooden structures. Violinists find that a wooden violin has a finite life and is eventually "played out" when the tone changes after long service. (A violin, of course, is subject to resonant loading.)

A measurement problem arises from the fact that most composites conduct heat poorly compared to metals, but their internal damping is higher. This means that the rate of cyclic loading must be limited so that the specimen does not heat internally to some meaningless value. Middle-aged materials scientists want results *now* and cannot wait around for a test machine to slowly grind out the many thousands of cycles needed to break the part without heating it. Even a meaningful definition of failure is hard to come by.

The author's conclusion, after reading papers on the subject, is that fatigue in the composite structures of interest to an amateur is not a significant factor. Propellers may be an exception, but *composite propeller design and testing should definitely be left to professionals, especially if there is any question of resonant loading.* Some current aircraft designs use composite landing gear struts which might raise a question of possible fatigue. Even at the very high rate of 2,000 landings per year, we may assume that few will be real tooth-rattlers and that it would take many years to get even near to a fatigue failure region. Even for this hard-service case, the gradual loss of stiffness near end the of life would be observable. Realistically, the struts would be replaced before catastrophic failure would occur.

Environmental Effects

Composites in the Boat Industry

We pilots have no doubt noticed that composite materials have been wildly successful in the small boat industry. It is the familiar boat application that most people think of when they consider "plastic" or "fiberglass" extended to aircraft. As used in boats, any serious saving in weight or material costs is secondary to the need to build a large, comparatively light weight, corrosion-resistant structure without leaking seams. Most boat manufacturers use methods roughly similar to those used in composite aircraft but have much larger production quantities that any aircraft company would envy.

There are very important differences: Boat manufactures are less concerned with weight than aircraft builders and are very much concerned with cost. If there is the slightest doubt about a part being able to carry a design load, they can casually use more material (at the cost of weight) to reduce the stress. They use much cheaper materials, often in the hands of semi-skilled workers who receive limited supervision. Engineering design of a boat rarely approaches the level of detail and test verification required by a good aircraft design. It is just easier to float than to fly, so we see some boats offered for sale to amateur buyers that are really poor structural designs, badly executed. Still, we observe that they usually run or sail to the satisfaction of their owners and they may sit out neglected in the hot sun on salt water for many years, with far less maintenance than the same design in wood. In fairness, there are many good boat designs which use the same basic family of low-cost fiberglass and matrix materials.

But we will agree that successful tolerance of the corrosion, radiation, temperature, and humidity of the marine environment is a major motivation for boat manufacturers to use "fiberglass plastic." A pilot might browse through any salt water yacht basin, looking at some obviously neglected boats, and ask herself how she would feel about storing her aluminum Bonanza or her husband's steel tube Citabria in such a place. She might conclude that a yacht basin is a bad place for either a yacht or an airplane, but that the fiberglass yachts don't do *too* badly, considering the conditions. Many ten and even twenty year old boats look very good. Some wooden trim may be pretty far gone at that age, but the wandering pilot would conclude that, even in that hostile environment, "fiberglass" works. If the technology works so well for boats, we ought to be able to build a well-engineered composite airplane structure that would last indefinitely.

Environmental Degradation Mechanisms

Pure Fibers in Water

The amateur airplane builder is not likely to encounter the really exotic composite materials as used in spacecraft, so we will discuss only those composites that use typical fibers in a plastic matrix.

First, let us consider some environmental effects on the fibers alone. By almost any standard, glass, Kevlar, and carbon fibers are extremely inert materials. We see that beer and wine are stored in glass for decades and that Kevlar 49 rope is used in extreme marine applications by the Navy and the oil drilling industry. (Carbon is too expensive and too brittle for such common use.) Materials scientists have tested these materials for extreme environmental limits with some surprising results.

Avenson, et al, in 1980 made some interesting tests on glass, Kevlar 49, and carbon fibers without a matrix. These materials were each held *under constant stress* in distilled water. With time *under stress*, glass and Kevlar lost strength and carbon was totally unaffected.

Beer drinkers will be surprised to learn that glass was the worst of the three. Glass displayed stress corrosion, leading to new cracks, analogous to the phenomenon we see in metals in a corrosive environment. The chemistry is complex, but glass held under stress in distilled water results in selective attack on the microstructure of the glass, similar to the selective attack on a crystalline metal in a corrosive environment. If an acid or alkaline environment is used instead of pure water, the effect is more rapid. Long term static stress in the presence of humidity is necessary for this effect to be seen.

Kevlar fiber can absorb water as much as 6% by weight with long term immersion and will have dimensional changes as a result. No realistic manufacturer will soak his Kevlar in water before doing a layup, but it is important for the composite designer to understand that Kevlar must be dry before use. It happens that water-soaked Kevlar fiber does not lose strength at ordinary temperatures. If used at high temperatures, wet Kevlar *does* lose strength, but here we are very far from practical aircraft applications.

We note, however, that we do not use naked fibers by themselves; in a composite, the fibers are protected from the environment by being immersed in the matrix.

Composites and Water

The resins used in aircraft construction have an open molecular structure that permits water molecules to diffuse into the matrix. The actual degree to which this happens depends greatly on the specific plastic being considered. With epoxies, the effect of water is to reduce the stiffness of the resin slightly and to produce small dimensional changes which produce residual stresses with the embedded fibers.

Dickenson, et al, in 1984 performed another very interesting experiment. They used three separate series of laminated composite specimens of carbon fiber, Kevlar, and glass, all in the same Code-69 epoxy resin. The specimens were tested (1) in the original state (dry), then (2) were boiled in water for three weeks and tested, then (3) were re-dried after the boiling and tested.

Allowing for experimental error, there was *no difference in strength* in the three cases for the carbon fiber or the Kevlar composite.

For the glass composite, it was found that the boiled specimen lost approximately half of its strength. The amazing result is that the specimens tested significantly *better* after they were re-dried than in the original dry state! The experimenters reasoned that there was only a physical bond (which didn't change) between the matrix and the carbon or Kevlar, while the glass had a reversible chemical bond which was broken by the boiling and re-established when dried. Most pilots would consider boiling an airplane for three weeks then drying it to be a little harsh, even if there was an improvement in properties at the end.

The result does point to the fact that moisture could theoretically have an effect in a glass composite structure in extreme conditions. This effect can be reversible, rather like in wood structures, where slow changes in normal water content can often be tolerated indefinitely, as with old pianos and violins. This suggests that a composite structure which is exposed to moderate wet-dry cycling (like a parked airplane) will not suffer significant accumulated damage over a long time period.

Wicking

An important function of the matrix is to protect the fibers and the bond to the fibers from the environment. To give this protection, the fibers must be completely immersed in the matrix, without exposed ends. In the discussion above, we considered that water molecules would enter the matrix only by diffusion from the surface. Engine cowlings, fairings, and access doors are often composite structures which have been laid up on a formed mold, trimmed with a hack saw, then attached to fittings by bolts in drilled holes. Drilling a hole or opening a saw cut line in a completed composite structure exposes the fibers and the fiber bond directly to the environment at the edge and the protection function of the matrix is reduced. Water or another foreign substance can attack the bond line between the matrix and the fiber and open the bond irreversibly. The damage can be caused by direct attack of the bond or by differential expansion between the fiber and the matrix which is sufficient to break the bond.

Obviously, if it is necessary to trim or drill a composite structure the best practice is to seal the cut edge with a compatible sealing material.

Composites and Corrosive Chemicals

Old gasoline station storage tanks made of steel are being widely replaced by new tanks of glass composites. The old steel tanks usually wore out because water from the air condensed and pooled in the bottom of the tank, rusting out the bottom. The gasoline didn't bother the steel, but the water on the bottom, or in the ground outside, did. The new glass/resin tanks are effectively immune to both attacks. We have seen that glass and some matrix materials are *not* absolutely unaffected by water, but they are close enough to it to make very successful practical boats and storage tanks.

The chemical industry uses composite tanks to store highly corrosive chemicals like sulfuric acid. Where the chemical to be stored is particularly active, the tank designer can specify matrix materials which have particularly low susceptibility to water or to the attacking chemical. Alternatively, the designer may choose to line the tank with a different higher cost resin which is resistant to the chemical. Most current composite aircraft designs use composite fuel tanks or wet wings and so must consider the possibility of attack by fuel. The Lancair uses an expensive fuel sealing compound. Randolph #912 Alcohol-resistant Sloshing Sealer is sold by Aircraft Spruce for both aluminum and fiberglass tanks. Tony Bingelis in his excellent *Sportplane Construction Techniques* does not recommend the use of any sealant because there are no seams or joints in a fiberglass tank anyway and we don't want particles of an added sealing compound breaking loose.

> *With fuel tanks, follow the designer's instructions. Don't substitute or add some other material that you think resists avgas or autogas just as well, without thorough testing!*

A factor in storage tank design is the fact that stresses are typically very low. As we discussed above, glass is subject to stress corrosion. In the storage tank application, the stress effect will usually be negligible.

A potentially important limitation on the use of composites in aircraft is the fact that most chemical paint strippers used to strip paint from aluminum aircraft cannot be used on composites. These metal strippers will directly attack most matrix materials. Special chemical strippers are available for composites.

> *The author was told of a case at an unnamed California Air Force base where military aircraft with composite panels were being stripped of paint with a chemical stripper. The panels were clearly stenciled with a message warning against the use of chemical strippers. The warning was hard to read when the panels were coated with the gelatinous stripper, so the warning was not understood, resulting in the damage and replacement of many expensive parts.*

Radiation

Most polymers are sensitive to high energy radiation because the chemical bonds that hold the long molecules together in a useful way can be broken by the radiation. The long-chain polymers can be cut, resulting in shorter-chain polymers and possible new cross-linking to nearby chain segments. The effect of the cutting is usually a lowering of the polymer viscosity, a reduction of the softening temperature, and a reduction of mechanical strength. New cross-linking may cause an *increase* in the strength and ductility up to some limit. Above the limit, continuing cross-linking can produce lower strength and greater brittleness.

In fact, some of the very early glass/plastic composites were intended to be catalyzed by direct ultraviolet radiation, generally by moving the fresh layup from the shaded workroom to the sunlight outdoors. The matrix remained a liquid until cross-linking occurred from the UV in the strong daylight.

In practice, the most common long-term effect of UV radiation is to cause brittleness of the matrix. Small dimensional changes can cause internal stress, which, added to any significant external stress, can cause rapid deterioration.

Some foam materials are particularly susceptible to UV attack and must be used in such a way that they are protected from UV.

Glass and carbon fibers are unaffected by indefinite exposure to natural UV. Carbon fiber is actually an effective UV shield for matrix material lying below the surface. Surface matrix material, of course, remains vulnerable. Dickson, et al, in 1984 ran a 2500 hour test in a Sun-Test cabinet of a Kevlar 49 and glass epoxy laminate and found the properties to be substantially unaffected, even though some darkening of the surface matrix took place.

The worst practical environment is the salt water yacht harbor, where the combination of ultraviolet radiation, wet/dry cycling, and daily temperature cycling eventually can cause deterioration of the surface. If a composite boat can survive this, so should an aircraft.

Clearly, the best thing to do is to keep ultraviolet radiation away from a composite surface. In the long term, a dry hangar is the best and fabric covers are the next best. This is just as true for metal or wood aircraft, of course, but the composite aircraft will not be subject to "hangar rash" corrosion which metal aircraft sometimes get due to humidity condensation in unheated hangars.

But even with a hangar or a cover, we must face the fact that airplanes spend a lot of time outside and a surface must be provided which will prevent significant penetration of ultraviolet radiation.

Good paints do this.

Chapter 4
Theory of the Structural Sandwich

From the 2-D Beam to the 3-D Core Panel

Reviewing Simple Beam Theory

In Chapter 2, we discussed how forces flow in an ideal "flat" beam. The purpose was to lead the reader from the two-dimensional beam to an intuitive feeling for another very important aircraft structure, the *structural sandwich* or *core panel*.

In our simple model, we used the *downward* force of weights at the end of the beam as the load to be carried. We saw how the horizontal forces in the tension and compression members of the beam *increased in proportion to the length of the beam*, but the diagonal forces *stayed constant*. The horizontal members carried the largest forces to a distant attachment point, but -being horizontal- could not directly lift the weight. All members in the ideal structure were absolutely essential to develop a force to oppose the fall of the weight.

A very important point of the above discussion is the fact that it was *essential to transfer shear forces across the beam*. In our ideal model, we had cross members at 45 degrees to transfer the shear forces. We briefly considered more complex practical designs to transfer the shear forces, such as wing spars which may use a solid wooden plank or an aluminum plate.

Stresses in a Three-Dimensional Structure

Suppose that we made a three-dimensional lattice girder by placing four of the ideal four-panel lattice girders we discussed in Chapter 2 in a row and connected every node of the lattice with a new diagonal brace to each neighboring node. (The result would be more complex than the rigging of a World War I airplane and would be impractical to draw.)

Suppose, instead, that we replace each of the four lattice girders with a *flat plate* of the same overall dimensions and connect these beams with plates wherever the original lattice had a node. The result is the sort of square "egg crate" sketched below. (For this idealization, we will ignore the fact that so thin a plate would fail by buckling wherever there was compression so we'll assume that the plates are "thick enough".)

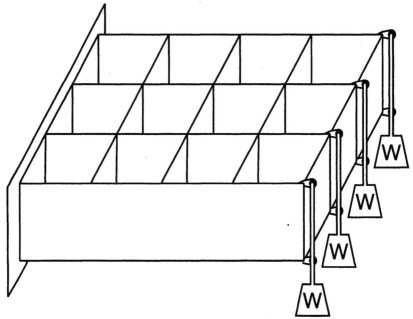

"Egg Crate" Beam - Evenly Loaded

The sketch shows four identical weights attached to the ends of the four beams. Each beam deflects the same, so there is no stress carried by the crosswise panels. The four beams carry their loads independently.

Now let us remove three of the weights, leaving the farthest weight in place, as shown on the next page.

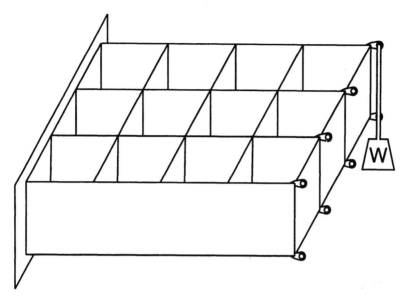

"Egg Crate" Beam - Off-center Load

Now the beams receive stress passed by the crosswise panels. In fact, if the loaded combination of beams is looked at from the end, we have something that looks much like our original four-panel beam with the load on the right end. The difference is that the total load is being passed back to the attachment at the wall in a complex way, with at least some stress being carried by every part of the structure. As with the simple flat beam, tension and compression loads are carried by material near the top and bottom of the panels and shear stresses are carried by material in the middle.

We haven't said anything yet about force components in a *sideways* direction. The structure sketched so far cannot take any lateral force component without collapsing. This can be verified by examining the cardboard "eggcrate" bottle divider from a case of wine. If the divider is removed from the box, it readily collapses. We would have a similar result with our ideal structure, as shown in the sketch on the following page.

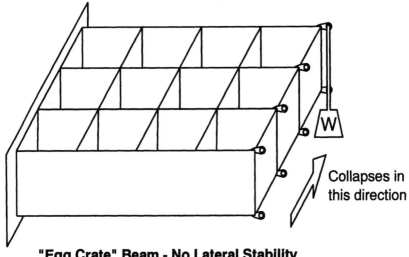

"Egg Crate" Beam - No Lateral Stability

So far, our ideal eggcrate structure has supported a load which acts only in a downward direction. A practical structure must carry lateral loads. One way to fix our structure would be to attach tension wires across the nodes, as shown below.

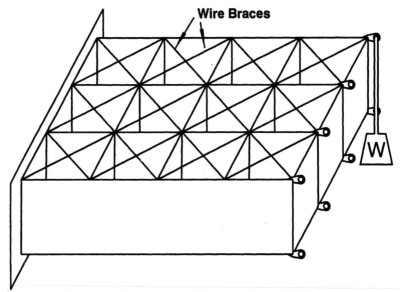

Braced "Egg Crate" Beam - Laterally Stable

Now any lateral force on the structure will place one set of tension wires under load while the remaining wires go slack. Shear stresses are carried by the vertical panels and the structure is stable.

Another way to stabilize the structure is to replace the network of crossed tension wires with a pair of solid plates, securely bonded to the beams, as sketched below.

We have just invented the structural sandwich!

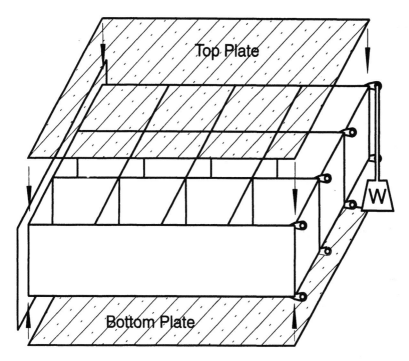

Derivation of the Structural Sandwich

The Structural Sandwich

Let's redraw our ideal eggcrate beam to show it with top and bottom plates of appreciable thickness, but still thin compared to the overall thickness of the beam. We'll put all the weights back so that the beam is evenly loaded across the section:

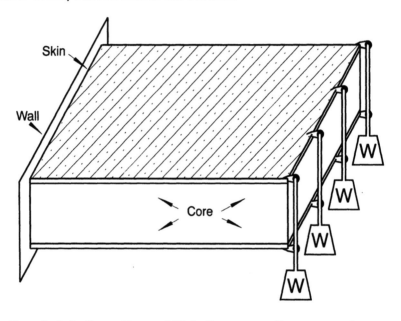

Sandwich Core Beam With Constant Cross-section

We might make a real beam much like this from thin plywood and sheet aluminum and we understand from Chapter 2 that the maximum loads will be at the *attachment* at the wall, where the top plate will be in tension and the lower plate will be in compression.

Bending

The sketch below is an *end-on* view of the loaded beam drawn previously. If we consider the stresses at the attachment to the wall, we see that the fibers in the top skin are in tension and the fibers of the bottom skin are in compression. The example is symmetrical, so the tension stress is equal to the compression stress. Without attempting to apply specific numbers, we show a constant stress across the entire section for the evenly loaded flat beam. This stress results from the load deforming (straining) the skin material. The stressed cross section essentially pivots around the center line, with the actual strain being proportional to the distance from the center line. (Thus, there is no tension or compression stress at the center plane. Engineers refer to the line of no tensile/compressive stress as the *neutral plane or neutral axis*.)

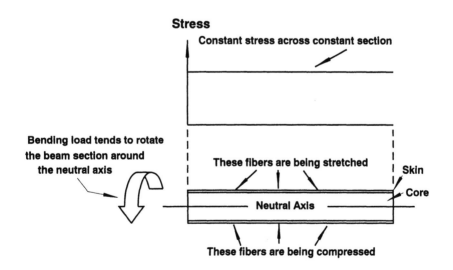

Section of Flat Sandwich Beam Under Load - End View

Failure Modes in Beams and Sandwiches

Suppose we made a beam or a sandwich structure, using two stiff skins with a core of less stiff material. This might be two sheets of steel bonded to a core of plywood or a standard all-steel "I" beam where the top and bottom are wide sections and the middle "shear web" is narrow. Let us mount this beam between two supports and apply a load at the middle, so that we see a considerable deflection, as sketched below.

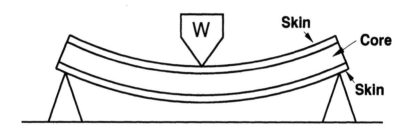

See that this arrangement is equivalent to two of our ideal cantilever beams, discussed in Chapter 2, connected back to back. The sketch below shows the same stress arrows we derived in Chapter 2 for the simple beam. In this case, the top skin carries compression

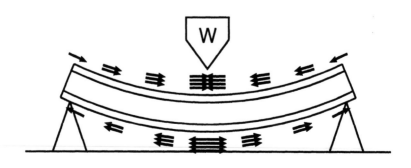

stress and the lower skin is in tension. Tension and compression stresses are at a maximum in the middle, as shown.

Core Failure

Let us increase the load on our test beam until a failure occurs. One failure mode would be extreme deformation of the core (crushing) with the skins left intact. Suppose we had tried a test beam with steel skins and a core of low grade expanded foam as is used in fast-food coffee cups. The failure might look like the sketch below.

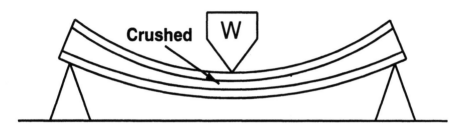

Tension Failure

Now suppose we kept the steel skins but used a better core - maybe plywood - and repeated the experiment. We might find a load so that the lower skin separated in tension, as shown on the next page, and the top skin and the core remained intact.

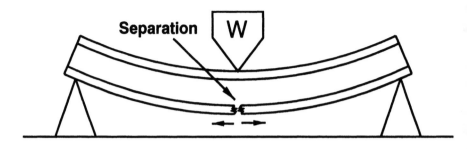

Compression Failure - Buckling

Alternatively, we might find the failure to first occur by crushing the top skin in compression. The sketch on the next page shows the case where a section of the top skin has sheared out of the plane of the skin. A variation would be to break the bond between the skin and the core, which allows the thin skin to buckle.

In practical sandwich structures, compression failure is usually the first to occur.

An example is the modern high performance composite ski, which rarely breaks, but, when it does, almost always buckles the skin.

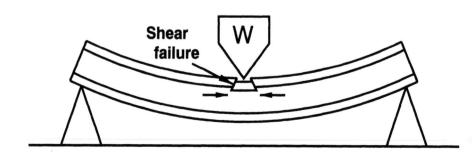

Stiffness in Practical Sandwich Structures

Most of the weight in a sandwich panel is concentrated in the surface facing skin. The core may be many times the volume of the facing skin, but is usually made of a very low density material whose function, like the diagonal members in our beam example, is to transfer shear forces over a large area between the facings. *The discussion above has emphasized that it is extremely important that the bond between the core and facing be very secure, so that the shear forces can be transmitted.* The core material is usually so light compared to the skin that it adds very little weight to a sandwich beam to make the beam thicker.

Hexel has done an interesting illustration of the effect of core thickness: if a reference beam is made of aluminum and then the aluminum is divided into two facing sheets and used in a sandwich, the stiffness is increased 7 times if the thickness is doubled and 37 times if the thickness is doubled again! In the comparison, the weight of the quadruple thickness sandwich is only increased 6% by the added core material.

So far, we have discussed sandwich structures which are flat, with parallel facings on a flat core, because we can get a general idea of how the forces must flow through the structures to carry the applied loads. Useful structures are generally more complex. In the *LongEze*, *Dragonfly*, and others, the control surfaces are foam cores cut to an airfoil shape and covered with a composite skin; there is no other internal structure, like spars or ribs. Large structures, like the wings, consist of large foam sections with integrated composite spars. Ribs, in the ordinary sense, are not used.

A way to look at the distribution of loads on a foam core wing is to compare it to the fabric covered wooden wings of World War I. The designer had the problem of getting the air loads which are distributed over the stretched fabric to the spars and - ultimately - to the fuselage attach points. If the designer were to double the rib spacing, stronger ribs and stronger fabric would be required to carry the doubled load. If the spacing were reduced to one inch, very light ribs would then be possible. The foam core is then seen as the limiting case of very close, lightly loaded ribs.

Flat sandwich panels are very fine for aircraft floors and bulkheads, but they are not useful for airfoil shapes. Let's show the principle with a similar beam, using the same constant skin thickness and the same core material at the same maximum thickness as the first flat beam, but now let's make the beam section two circular arcs, so that the edges are sharp. See the sketch below:

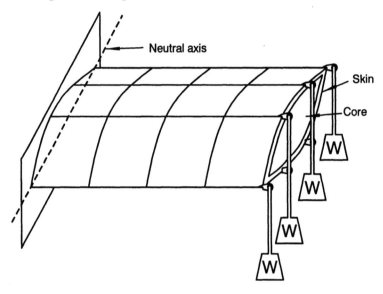

Sandwich Core Beam With Curved Cross-section

The sketch is intended to suggest that this beam, like the previous case, is evenly loaded across the section. Further, we'll assume that the total load is adjusted so that the *deflection under load is the same as the previous case.*

The drawing below shows the bending stress at the attachment for this case:

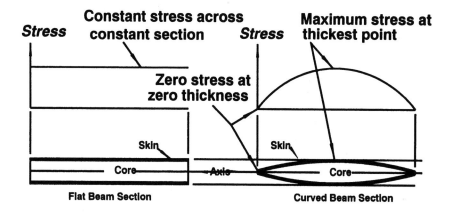

Stresses in Thin-skin Sections

See that the new curved-section beam cannot carry the same load as the flat beam at the same maximum fiber stress because most of the skin material is closer to the neutral plane and does not receive the same strain as at the widest point. In fact, *the edges of the beam carry zero load.* In the sketch, we show the stress in the skin as proportional to the distance from the neutral axis. The total load carried for this case would be less than half that for the flat beam case. The sketch on the next page shows a comparison of the stresses at the attachment for the flat beam and the curved beam.

Now let's look at another curved sandwich beam where we *double* the thickness of the reference flat beam, as sketched on the following page.

Now the maximum stress for that deflection has doubled because we have doubled the core thickness. The load carried is *more* than double that of the flat beam, however, in spite of the less-effective edges. We could carry the same load as the flat beam at the same maximum stress at about half the deflection. This thicker beam is much stiffer even though the amount of skin material is only slightly increased. The core volume is increased considerably, but we note that for aircraft applications most of the weight of a sandwich structure is concentrated in the stressed skin.

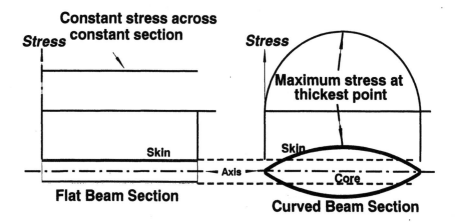

**Stresses in Thin-skin Beam Sections
At the Same Beam Deflection**

The engineering theory of beams shows that, for a given amount of material and weight, a thick beam may be many times stiffer than a similar thin beam. We know that a board loaded edgewise is much stiffer than the same board loaded on the flat. (In fact, other things being equal, the contribution of any stressed element of a beam to the stiffness of the beam is proportional to the *square* of the distance of the element from the neutral axis.) *Conclusion:* thick beams are much more efficient than thin beams in utilization of material to carry a load, so a thick structural sandwich is very much stronger for its weight than a thin one.

Complex Loads

So far, we have used simple beam theory to discuss only the simple case of transferring a single ideal load from a single point of application to another place. Practical structures must support complex combinations of loads which the creative structural engineer must analyze if he is to design a structure to carry the real load.

A uniform load is one in which the load is the same for every area of the supporting structure. Examples are compressed air in a tank, water in an aqueduct, or a continuous line of people standing on a bridge.

Air Loads On A Wing

A distributed load is one in which the load varies in some continuous way over the surface of the supporting structure, without being concentrated at specific points. An example is the distribution of lift over an airplane wing, which is applied continuously to the skin of the wing - positive pressure on the lower surface and negative pressure on the upper surface. This distribution is quite complex, varying from nearly zero at the wing tips and near the trailing edge to some maximum near the middle of the wing. Close knowledge of this load distribution is the specialty of professional aerodynamicists. The structural engineer has the task of matching the structural design to the actual load distribution. Passengers on current jet transports look out the window to see that the center third of the upper surface of the wing often appears to be a big smooth plank of different colored metal; the wing designers have put the strong material where the stress is.

The fact that flight loads are near zero at wing tips allows aircraft like the Piper *Cherokee* and Beech *Sundowner* to use wing tip fairings of lightweight unsupported and unreinforced molded plastic which are attached to the sturdy metal wing with a line of small machine screws. This suggests that the metal wing is grossly overdesigned (and overweight) for some distance inboard from the fairing. Considerations of manufacturing simplicity with a skin of uniform aluminum sheet force a design which is too heavy outboard so that it is strong enough inboard. In principle, a design using composite materials can more conveniently allow a tapering of material to support a tapering load.

"Handling" Loads

Only a few areas in an airplane design are really design-critical for flight loads. These would include such obvious parts as wing spars and skins, landing gear and wing attachments, load spaces, and fuel tank mountings. Much of the surface of an aircraft consists of fairings and low-stress areas which see very low flight loads but must stand up to pushing, occasional flying gravel, feet that ignore the "no step" sign, and the other bumps and bruises of daily life. As a practical matter, these areas must end up overstrength for flight loads because handling loads are real.

Chapter 5
Structural Fibers and Matrices

Fiber Choices and Fiber Bonding in Composite Structures

In Chapter 3 we discussed why it is that we use the "strong" component of a composite structure in the form of fibers: Essentially, we found that we can only get these very high strengths in small sections of the material - inevitable flaws occur and limit the strength of large sections. We showed that much of the inherent strength of the "strong" component could be obtained in a large finished part, such as a wing spar, if the flawed fibers were immersed in a less-stiff matrix which passed load around the flaws, from one fiber to the next. In this chapter, we will discuss the characteristics of specific fibers that the home builder will be likely to use and some details of the bonding between the fibers and practical matrices.

It is a reality of life that the leading designers of aircraft and aircraft kits for the homebuilt market cannot afford to be the pioneers in materials development. Exotic new materials first appear in military/aerospace applications, then appear in expensive sporting goods such as golf clubs, tennis rackets, and bicycle frames. Finally, when sufficient industry volume develops, the cost becomes acceptable for the home aircraft builder. The designer of home built aircraft will then adopt a system of fiber reinforcement, matrix, and fabrication techniques which has been proven by some successful pioneer. (The best example of this is the contribution of Burt Rutan's 'Eze series to a number of derived designs, like the *Dragonfly* and the *Cozy*, which used Rutan's materials and methods.)

Glass Fiber

The manufacture of "fiberglass" is a large and competitive world industry. Huge quantities are made in wide variety for building insulation, air filters, molded plastic reinforcement, and many other industrial uses. Glass fiber is manufactured by ejecting molten glass through tiny holes in a platinum die, much like a spider ejecting web material, then winding the cooled filaments on a spindle. The raw material is usually common window glass, so is very cheap.

Glass fiber filaments that would be used in a structural application are uniformly quite small, well under one thousandth of an inch in diameter. Filament sizes range from 0.00013 inch (B fibers) to 0.00051 inch (K fibers). Individual filaments are grouped into *strands* of as few as 51 filaments to as many as 2052, depending on the application and the details of manufacture. Usually, all the filaments making up a strand are taken from one pot of molten glass at the same time and are called a *"tow"*. Tows of glass fiber may be twisted together, then twisted with other strands to make *yarn*. Yarn is then delivered to weavers who manufacture the many glass fabrics that are used in construction of actual parts.

A *roving* is a single bundle of parallel untwisted fiberglass strands which is used in the manufacture of stitch-bonded fabrics (see below). Roving size is expressed in *yield*, which is the linear yards per pound of a single bundle.

The most common glass material by far is designated *E-glass*, which has outstanding physical characteristics at the lowest price of all the structural fibers. For E-glass, best tensile strength of the fibers is about 500,000 psi, and modulus of elasticity (E) is about 10,500 kpsi @ 72 °F.

Several other grades of raw glass material are available for commercial glass fibers, but only one other, *S-glass*, is used in aircraft construction. S-glass has a tensile strength of about 665,000 psi and E is about 12,400 kpsi, so S-glass is roughly 30% stronger and 15% stiffer than E-glass, at about triple the cost. It also retains strength better at higher temperatures, though this may not be important if the final system is limited otherwise by practical matrix materials. Rutan's *Defiant* specifies S-glass strand material. *S-2 glass*, is a variant of S-glass made by Owens Corning which has somewhat better properties. A very interesting form of S-2 glass is *Hollex*, also manufactured by Owens Corning, which is composed of hollow fibers. The resulting fibers have approximately the properties of E-glass, but with 30% less weight.

Carbon Fiber

Carbon fiber is a good example of a one-time prohibitively-priced "exotic" material which has been brought from the cost-no-object aerospace field, through the golf club and tennis racket stage, and into the range of affordability of homebuilt aircraft. Physical properties are indeed excellent: Strength = 470,000 psi, E = 34,000 kpsi, at less weight than glass. The very best sailboat masts are made of carbon. The round-the-world *Voyager* was made almost entirely of carbon. The pressurized *Lancair IV P* is made of carbon (but the lower-priced fixed-gear *Lancair ES* uses E-glass). Leading homebuilt aircraft designers find that carbon is now a good buy for strength, weight, and cost, and use it in a few critical areas like spar caps, while using inexpensive E-glass for the rest of the structure.

Carbon fiber composites have a rather nasty handling characteristic that the builder should be aware of: Breakage of a structure gives rise to hundreds of shattered needle ends that have a remarkable tendency to get into the skin and cause irritation. Breathing of dust from sawing *must* be avoided, too.

Kevlar™

Kevlar aramid fiber, a product of the DuPont Company has the highest strength-to-weight ratio in tension of any commercially available fiber. It has been extremely successful in such varied applications as premium tire cords, marine cordage, military body armor, propellor blades, and high pressure rocket casings and oxygen bottles. It has found wide use in commercial and military jet aircraft for a variety of applications. It is 43% lighter than glass and around 20% lighter than most available carbon fibers. Three formulations are available: (1) Kevlar 49, usually used for structures, (2) Kevlar 29, used for as a protective lamination with other materials and where superior resistance to local puncture is desired, and (3) Kevlar 129, used for body armor.

Kevlar has outstanding properties under tensile load, (strength = 430,000 psi and E = 19,000 kpsi) but shows very much lower useful strength in compression, so it is not used for spar caps. It does find use in engine cowlings and in wheel pants (which are subject to damage from flying gravel).

The material may be somewhat of an anomaly when considered from the viewpoint of our discussion in Chapter 3. With glass fibers, we assumed that there would be many flaws which would effectively break a single glass fiber into short segments. With Kevlar, there evidently aren't many such flaws, so Kevlar fibers have very long unbroken segments and are not very dependent on the bond to the matrix in order to obtain useful tensile strength. Available matrix materials do not bond to Kevlar nearly as well as to glass or carbon and this may account for the measured poor compression strength of Kevlar composite structures.

Experiments have been reported in which Kevlar fibers have actually been lubricated so that there will be only negligible bonding to the matrix, then laid up and tested. The lubricated specimens actually tested slightly stronger in tension than the corresponding bonded specimens!

Kevlar is extraordinarily tough, so tough that special tools and techniques are required to work the material. It does not sand well, but produces many little whiskers on the surface. *Precautions must be taken against breathing Kevlar dust.* A surface layer for sanding, usually glass, is sometimes specified for Kevlar structures.

Ceramics

Mineral fibers of about the same mechanical properties as S-glass are available which can be woven into cloth. These fibers have the property of being able to stand very high temperatures, so are finding use as fire protection barriers. The disadvantages are extremely high cost and the fact that few matrix materials come anywhere near a comparable temperature tolerance.

Hybrids

Fibers of different characteristics can be mixed to produce new composites with very interesting properties. KS-400 is a blend of Kevlar 49 and Owens-Corning Fiberglass S-2 glass which is intended for hand layups. The glass component facilitates wet-out and the very good matrix bond to the glass provides excellent resistance to delamination.

Other combinations can be used. Carbon fiber, used alone, tends to produce a brittle structure. A corresponding S-glass part is much tougher, but is not as strong as the carbon part. A hybrid S-glass/carbon part has nearly the best performance of each material used alone. Various mixtures of S-glass, carbon, and Kevlar are used in a number of structures in late model Boeings. At this writing, only the KS-400 combination, offered by Aircraft Spruce, is available to the home builder community, but technology progresses quickly!

Bonding to the Matrix

Clearly, a useful two-phase composite material depends on the integrity of the fiber-matrix bond, whatever the inherent strength of the matrix material itself. Material manufacturers and fabricators have given much attention to this subject and their recommendations for handling and processing should be followed very carefully by the home builder. It is necessary that the bond be stable for the life of the structure and not subject to attack by the environment or to deterioration over time.

There are two conflicting requirements in the manufacture of useful fiber materials. The first is that the material must he handled during manufacture (as in during weaving into a fabric), which usually requires that some form of lubrication be applied. The second is that a preparatory coat may be required to assure that a permanent bond is formed with the matrix.

In the case of glass fiber, a "binder" made of starch and oil is applied to the raw tow. This "binder" (actually a lubricant) facilitates the weaving process, but must be removed before the appropriate "finish", or coupling agent, can be applied. The cleaning is done by baking the woven goods in a hot oven.

After cleaning, any of several coupling agents can be applied, most being totally unsuitable for aircraft use, so it is essential to assure that the appropriate "finish" has been applied. One of the best, and suitable for all of the generally used matrix plastics, is "Volan A". If clean glass fiber is laid up in a matrix without the use of a proper "finish", water will run along the interface between the glass and the plastic and will destroy the bond over a period of months.

> *For aircraft construction, then, it is imperative that glass be known to be clean and freshly treated with the finish appropriate to the matrix to be used. This rules out unknown "fiberglass" from auto body shops and paint stores. If it is to be flown, obtain clean, fresh material from an aircraft supply house and keep it protected until use.*

Carbon "tow" is difficult to handle in weaving and cannot be baked or washed to clean, so the necessary lubrication is provided by adding a small amount of epoxy to the surface. This epoxy remains on the finished fiber and must bond with the matrix to be used in the finished product. Again, manufacturer's recommendations should be followed.

Sizing and binders are also used to facilitate handling of Kevlar. Kevlar cannot be cleaned in an oven, so it is essentially washed in water to remove these coatings. No coupling agent has been found which produces a bond between Kevlar and a resin which is as good as a glass-resin bond. This bond remains a limitation in designs using Kevlar.

Fabrics

It would be nice if we could place all of the structural fibers to be used in a structure exactly where we determined that the load forces would flow. The millions of tiny fibers would be aligned with the forces only where needed and we would have a structure of remarkable efficiency.

We would also have a design that would be practically impossible to build. We cannot handle and assemble a finished structure from a mess of loose fibers and liquid plastic. We need to secure the chosen fibers in an orderly way so that they are aligned sufficiently accurately with the loads and can be handled in reasonably sized sections during assembly. This requirement is met by weaving the fibers into a very wide range of cloth products and a very highly developed industry exists to provide this service.

Tow

"Tow" is the term used in the weaving trade for the basic bundles of fiber filaments that weavers use to make the many types of fabrics. Aircraft Spruce offers unwoven carbon tow on spools in three sizes. The home builder may find small quantities of carbon tow useful in such applications as distributing concentrated stress from a steel bolt to a surrounding glass structure through a bushing.

> *The reader will remember the introductory puzzle in Chapter 2 and will recognize that the stiff carbon will take nearly all the load from the adjacent less-stiff glass. Don't guess about the consequences of joining dissimilar strong materials. Don't "make it stronger" by adding a little carbon to a primarily glass structure unless you or the designer have actually tested the design.*

Unidirectional "Fabrics"

Spar caps, wing skins, and similar parts of an aircraft carry simple longitudinal loads. The designer of a composite structure would prefer to place the load-bearing material very nearly along the lines of stress. It is perhaps not entirely correct to call such a material a "fabric", but weavers furnish sheets and tapes of fibers which have nearly all of the fibers running only in the lengthwise direction. Unidirectional cloth consists of flattened bundles of fibers which are restrained by a minimum of crosswise stitching. The idea is to provide just enough crosswise material to make handling practical. The fact that the fibers lie smoothly next to each other produces an efficient layup with small gaps between fibers to be filled by the matrix. High fiber to matrix ratios result and nearly full theoretical strength is maintained. Of course, crosswise strength is considered to be zero because there is a minimum of load-bearing fibers in the crosswise direction.

Most current plans and kits for composite aircraft specify substantial quantities of unidirectional cloth for heavily loaded areas, like spar caps.

Unidirectional "cloth" is offered for all of the fibers used by the home builder.

Weaves - Bi-directional Cloth

The weaving industry offers a wide range of weights and weave types, particularly in glass. Remember that we use cloth because we need to handle and direct the fine fiber material efficiently and we accept some compromises to be practical. All single-ply bi-directional fabrics offered for the homebuilder have an equal number of fibers in the lengthwise (warp) direction as they do in the crosswise (fill) direction, hence the term bi-directional. In weaving, we give away a little of the theoretical fiber strength because the strands are twisted and bent; they must go around each other. This necessary curve is called the *crimp*.

Fiber quantity is expressed in terms of weight per area, or ounces per square yard.

A *plain weave* has every strand going alternately over and under the crossing strands. The fine cloth in a man's dress shirt or a coarse gunny sack has a plain weave. Plain weave materials have the characteristic that they have a limited ability to form to irregular surfaces; i.e., they will not easily follow small rises or depressions and prefer to remain flat, like a sheet of cardboard.

A *twill* or *satin* weave has one yarn crossing several yarns before going under the next single yarn. The weaver's trade has several very old terms for the several weaves, such as *twill, crowfoot,* or *satin,* depending on the number of yarn crossings These weaves have the useful property of being able to conform smoothly to moderately concave or convex surfaces. They have a little less crimp than a plain weave, so they should lie slightly closer in adjacent plies and give a small advantage in theoretical strength. A medium weight (8 - 9 oz/yd^2) crowfoot weave is probably the most versatile general purpose glass cloth that the builder will want to keep available in his shop.

Other Weaves

Knitted Fibers - Triax and Double Bias

Current composite designs usually call for skins to be composed of more than one layer of fiber, with two or more layers consisting of unidirectional fiber and at least one layer with the fibers oriented at +/- 45°. In older designs, the builder met this requirement by applying each layer of fiber in turn, as cloth, until the required number of layers was in place. The +/- 45° fibers are applied as woven cloth which has been cut from a standard roll at at 45°.

Triax is composed of three separate layers of unidirectional fiber which has been knit together by a knitting specialist with just enough supplementary fiber to hold the assembly together for adequate handling. The most useful configuration is one layer at plus 45°, one layer at minus 45°, and the third layer at zero degrees (lengthwise). See below.

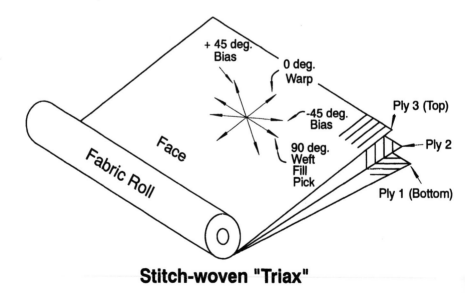

Stitch-woven "Triax"

This system has the considerable advantage that all three layers can be applied at one time, while keeping the fiber orientation accurate. Handling is as easy or easier than with conventional woven cloth and, like satin weave cloth, it conforms well to moderate curves. Of course, the ability to apply three layers at one time is a significant work saving. The finished structure is a little more efficient than the same layup done with cloth because most of the fibers in the layup are parallel, not "crimped", as with woven cloth.

Triax has only recently become available in the small quantities required by the typical homebuilder, but the suppliers now offer it in their catalogs. This material is so convenient to use that it will probably increase in use, even though it is somewhat more expensive than single-layer material purchased separately. Waste is reduced because the 45° layers are accurately trimmed with the edge of the roll and the knitted layers hold their form for cutting better than most cloth.

A significant advantage of these stitch-bonded materials is that the finished surface can be much smoother than if woven cloth had been used, because the fibers lie parallel. This reduces the need for a surface fill to get a smooth finish. The manufacturers use the term *gauge* to express the tightness of stitching. A larger gauge number means that there are more stitches per inch, which closes up the fabric and gives a smoother surface. The resulting tight fiber bundles produce a stiffer fabric which is harder to wet out and harder to conform to an irregular surface. A lower gauge allows the fibers to open more and conform better.

The ease of surface finishing has made triax attractive to manufacturers of snowboards.

Double bias knitted material is also available to the industrial composite industry. This is a two-layer material with the separate layers at +/- 45°. It is functionally similar to woven cloth, except that it is slightly more efficient in placement because it has zero crimp. It is also convenient to use because it comes off the roll with the fibers at the desired 45° angle. Like satin weave and crowfoot weave cloth, it conforms well to curved surfaces. Double bias knitted material is not currently available to the individual, but is of possible benefit to kit manufacturers and could be available to individuals if demand develops.

Knytex, of New Braunfels, TX (a joint venture between the Hexcel and Owens Corning Corporations) and Brunswick Technologies, of Brunswick, ME, supply a wide variety of special stitch-bonded materials to the aircraft, sporting goods, and ski industries.

Mat

"Mat" is not really a weave, strictly speaking. Mat is essentially a non-woven felt made of short lengths of fiber at random orientations. It is usually much thicker than a woven cloth of the same weight, so has larger voids than a true cloth. It has the considerable advantage of conforming readily to a curved surface to serve as a core in a sandwich structure. The main disadvantage is that a considerable quantity of resin is necessary to fill the voids, so fully-impregnated mat is rather heavy, compared to other potential core materials. Mat finds extensive application in boat building and auto body shops, where extra core weight is secondary to ease of use and cost of production.

In aircraft construction, mat is often used for mold-making because it is easy to get a relatively thick (thus stiff) structure with a single layup. It is useful for some aircraft repairs where the work of duplicating an original core structure isn't worth the weight saving for a small patch.

Adding Resin to the Fiber

In wet layup composite construction, the builder places a dense assembly of strong fibers into some desired final shape and then saturates that assembly with a liquid which hardens into a solid matrix. The ideal situation would be where the fibers were covered with a minimum of liquid matrix, the structure had no resin-free voids, and there were no areas of excess resin. To achieve this ideal, a liquid resin would have to have low enough viscosity to penetrate and wet a thick mat of closely-packed dry fibers, but be viscous (and "sticky") enough to coat the fibers, stay put, and not run down to low points and form pools of excess liquid. Practical liquid resins don't quite meet this ideal, but the techniques discussed in Chapter 10 show how to work around the difficulties.

The composites industry has other ways to add resin to a structure and a brief consideration of some of them will be of interest.

Pre-impregnated Fabrics - "Prepregs"

An alternative to adding liquid resin to fabric in the final form is to add the activated resin to the uncut fabric before assembly and to delay hardening of the resin by keeping the material under refrigeration. This technique is not practical for the homebuilder, but is very important for kit manufacturers who can make nearly complete parts made from these materials in precision tools. We will have much to say about this process in Chapter 11.

Manufacturers, called "prepreggers", buy woven fabric goods in bulk from weavers and apply activated resin to the fabric with great precision. The product is delivered in rolls with a thin sheet of plastic used as a separator on both sides. The fabricator then cuts the cloth to a desired shape with the plastic separators in place. The bottom sheet is peeled off, the cut cloth is placed on the assembly, then the top protective sheet is removed. The matrix is not liquid, but has a sticky surface about like grape jelly, so there is no tendency to flow down vertical surfaces and the cloth can be placed very accurately. The matrix content is closely controlled, so that the fiber/matrix ratio is higher than can be achieved with a hand layup using liquid resin.

The resins used in "prepregs" are hardened at high temperature in an oven, so the room temperature working time is of the order of several days. Complex assemblies of many layers and pieces may be accurately assembled on a mold, then heated to finish. Most fabric used by the commercial aircraft industry is, in fact, supplied as prepreg. Prepreg material is expensive, but handling is so easy that there is usually very little waste. The combination of superior performance, economy of labor due to the easy handling, and low material waste makes the use of prepreg materials very attractive to properly equipped composite manufacturers.

Other Methods

While not used directly by the homebuilder, some other methods of bringing fibers and resins together will be of interest. The day may soon come when aircraft parts made by these methods will appear in the General Aviation market.

Resin-transfer Molding - RTM

In RTM, a previously formed fiber structure (preform) is placed in an empty metal mold. The activated resin is pumped in to soak the preform and fill the mold. The special resins used for this process are usually quick-curing two-part materials and are mixed just before injection. The process can produce large, complex parts which are essentially finished, so has found wide use in the sporting goods industry. The automobile industry has used the process for many years and Boeing now uses it for production parts.

Thermoplastics

Those of us who are familiar with cheap plastic milk bottles may be surprised to learn that really excellent thermoplastic matrix materials are coming into use. A thermoplastic is one which becomes liquid when heated to some critical temperature and has good mechanical properties at the temperature of use. (Seen this way, cast aluminum is a thermoplastic.) Thermoplastic prepregs are available, so excellent complex finished parts with very high fiber density can be essentially "cast". The melting temperature of these thermoplastics is far above the temperature likely to be encountered in a practical application, so they are appearing in military aircraft.

Types of Resins

The home builder will work with resins which are shipped and stored as liquids. The room temperature hardening process is started by the builder just before use.

There are two basic types of room-temperature hardening resins used by home builders: (1) two-part resins where two accurately measured reacting components are thoroughly mixed together, and (2) an inhibited resin where the user "inhibits the inhibitor" to start the hardening process.

For Type (1) - the two-part system - it is very important that the reacting components be accurately measured if the final part is to reach full strength. An error in measurement will appear as an excess of one unreacted component which dilutes and weakens the final solid material. For Type (2), the system is tolerant to errors in measuring the small amount of activator because errors mostly effect the hardening time of an already correct mixture, not the final hardness.

In the discussion below, we will consider two-part resins and inhibited resins as separate groups and will make no attempt to teach organic chemistry or to make the reader into an organic materials scientist. We will use the product names given by the manufacturers and will avoid the chemical names.

The range of matrix materials available to the professional composites industry today is mind-boggling and the choice is certainly best left to the professionals. This is doubly true when choosing a "system" of fabric, core material, and matrix. If the reader is a builder of a proven design, then he should stick exactly to the tested system chosen by the designer. If the reader is designing a new airplane, and is also considering a new composite "system", then he should obtain professional advice and should test extensively if departing from the materials family used by a successful predecessor. We will discuss only those resins in current use in popular designs, and currently available from the aircraft supply houses.

It is appropriate to remind the reader that the composites field is moving rapidly and that many improvements in materials and methods have been made since Ken Rand and Burt Rutan started things and improvements continue to be made. The specific products described below may be replaced by better ones tomorrow.

Post Curing - The Glassy State

A crystal is an orderly array of atoms. Table salt makes perfect cubes of stacked sodium and chlorine atoms. Diamond is a crystal of interlocked carbon atoms. Steel is a network of crystals. Frozen water ice is a network of crystals. When ice melts, the water molecules leave the crystalline structure and are free to move away as liquid water. The transition between crystalline ice and liquid water is quite clear.

Quartz makes large crystals of silicon and oxygen. Crushed quartz, as sand, is a major component of ordinary glass. As glass, the material does not form accurately laid out crystals, but has the atoms only approximately in the crystalline relationship. A consequence is that glass does not have a sharp melting point like ice, but goes through an intermediate softening stage. Glass blowers make use of this intermediate condition. This characteristic of having a softening point between the solid and liquid states is so characteristic of glass that materials scientists refer to it as the *temperature of glass transition*, T_g, when referring to *any* material, not just glass.

There are quite a number of other materials which have a semicrystalline, or glass-like softening characteristic. Taffy candy is an example. *Cured epoxy is another.*

As a rule of thumb, the epoxies used in aircraft construction have the property that the softening point at elevated temperature, or the glass transition point, T_g, is about 40°F above the cure temperature, up to some maximum. This means that an aircraft wing laid up in an 80°F shop and put in the sun immediately in Phoenix in August is at risk of softening and distorting when it reaches 120°F. In practice, if the structure is heated slowly above the original cure temperature, additional linking occurs in the epoxy over time and the T_g moves up, staying ahead of the cure temperature by the same 40°F.

We see that epoxy aircraft are actually parked in the sun in Phoenix without problems and conclude that many get a successful post cure treatment, whether intended or not.

The best practice is for the builder to include a thorough post cure as part of the building process. It is most convenient to do this before the aircraft is assembled. A temporary wood frame curing house can be built and covered with commercial wall insulation. The parts can be held in place so that they are not under load and so that air can flow freely past the parts. Electric heaters can provide the heat for this small, well insulated space, and fans should be used to assure that the temperature is uniform.

Best practice is to cure at the highest temperature that the limiting structural material will allow. Usually, the limit will be the foam used, not the epoxy. Sources vary on exactly what the limit is for the foam materials used. Rutan, in his 1980 *LongEze* manual, refersto the practice of sailplane manufacturers to "putting the entire airplane in an oven at 160°F" and says that "blue foam is damaged above 240°F." Later sources say that blue styrofoam may be heated to 165°F. and that urethane can stand as high as 180°F.

> *As should be the rule in aircraft practice, if there is any doubt about the critical characteristics of any material, check with your designer and the specific source of the materials you are actually going to use.*

Rutan says that a completed structure, like a wing, be post-cured by simply painting it with black primer and standing it in the sun for a day. Other sources suggest that black water-soluble poster paint will do the job and be easy to wash off when the operation is complete.

Rutan also suggests that the post-cure operation is an opportunity to correct small errors in wing form. With the wing firmly held in position by clamps or Bondo, the wing is forced to the correct position and allowed to soften and post-cure. If this correction is done during the post-cure of both the top and bottom surfaces, the error will be corrected and the finished part will be stable for future loads at high temperature.

As a practical matter, post curing in a temporary oven at for about two hours at 140°F should produce a quite safe structure, not stress any structural materials, and be reasonably easy for the home builder to achieve. This treatment would produce a T_g well above the limitations imposed by foam materials.

> It is only a coincidence that we discuss a "glass transition temperature" when we are talking about the matrix, NOT the fibers. This point has nothing to do with the fact that we may also happen to use glass for the structural fiber.

Two-part Resins

Measuring Two-part Resins

It is necessary to measure the reacting components of two-part resins very accurately, otherwise an unbalance appears as a dilutant of unreacted material, reducing the strength of the final matrix. All of the plans for homebuilt designs that use two-part resins (that the author has seen) show details of how to make a precision ratio scale for accurately weighing the components. The author has observed several builders at work with two-part systems and offers the following suggestion:

> Working with matrix resins and loose fiberglass is a messy business and anything that makes it easier and saves expensive resin is well worth the cost. Repeatedly weighing small batches of resin on a scale is OK for a project like an auxiliary gas tank or a wing tip repair, but will get mighty tiresome and wasteful for an entire aircraft project. The author strongly recommends that a precision ratio pump be purchased which is appropriate for the resin system chosen. When resin mix is needed in the shop, the builder - sticky hands and all - just strokes the pump handle and the right ratio of components flows into the mixing cup.

Burt Rutan's VariEze - Some History

Burt Rutan was not the first to apply composite techniques to a homebuilt aircraft, but he must be credited with being the first to demonstrate outstanding success with a radical design that was entirely composite, not a hybrid, and delivering a complete technique to the amateur. He was indeed a pioneer designer who had to work directly with manufacturers to make available materials that were not previously sold to the small user. Through his efforts, the special cloth weaves, special epoxy resins, and new techniques that were necessary for the *VariEze* became available and were picked up by other designers. He started an entirely new branch of homebuilt aviation. Experimental Aviation owes him much.

The RAE Epoxy System

Rutan's original VariEze and LongEze designs specified an epoxy system manufactured by Hexcel which consisted of one component called the "resin" and a second component called the "hardener". *(These formulations are no longer recommended, but are described here for historical interest.)* The "hardener" came in two types: (1) Rutan Aircraft Epoxy Fast (RAEF), which had a short pot life, and (2) Rutan Aircraft Epoxy Slow (RAES) which had a longer pot life. RAEF had a pot life of 20 to 45 minutes and cured in 3 to 6 hours at 77 °F. RAES had a pot life of 1 to 2 hours and cured in 10 to 16 hours at 77 °F. Both formulations were mixed at 4 parts of resin to one part of hardener, by volume.

Both formulations produced good wetting, good handling properties, and excellent physical qualities in the finished part. RAEF was used for most applications in constructing an aircraft, where the compromise between pot life and curing time is practical for making smaller parts. RAES was used when longer pot life is required, as when laying up a large part like a wing. A problem appeared when a sizable minority of builders developed an allergic sensitivity to these materials - some quite severe - even though recommended precautions for use were followed. These resins built a lot of excellent aircraft and were still offered by aircraft suppliers until recently.

The Epolite (Safe-T-Poxy) Family

In response to the problem of allergic sensitivity to the Rutan formulations, a new epoxy system was developed by Hexcel and sold under the name Safe-T-Poxy, later renamed *Epolite*. This product produced physical qualities as good as the original with greatly reduced allergic problems and soon found wide use in later designs that used Rutan's methods. Like the Rutan formulations, the base resin (#2410) is used with appropriate hardeners to give a range of pot life and curing time. Three hardener formulations are offered: Standard (#2183), Fast (#2182), and Slow (#2187). The mix ratio for all three is 100 parts resin to 45 parts hardener, by volume.

The original Epolite/Safe-T-Poxy had the disadvantage that the viscosity at ordinary workroom temperatures was a little high and builders found some difficulty in getting good fabric wetting in multi-layer layups. A lower viscosity product named Safe-T-Poxy II was developed which had better wetting characteristics than the original with the same degree of safety. The better wetting was important when working with carbon fiber or Kevlar laminates. This system has a mix ratio of 7 parts resin to 3 parts hardener, by volume, a difference of about 5% compared with the original system.

Epolite/Safe-T-Poxy was a great improvement in reducing toxicity, but these materials still cannot be considered to be without danger. The Occupational Safety and Health Administration identified a component of Epolite called MDA as being a dangerous chemical. Hexcel subsequently introduced a replacement resin (#2427A) and a new hardener (#2427B) which contains no MDA and actually produces improved handling and physical characteristics. At 77 °F, pot life is 100 minutes, tack-free time is 4 hours, and cure time is 10 hours. The material may also be cured at 150 °F, in 2 hours. The new 2427A/B system is now recommended for the *Dragonfly*.

Epolite/Safe-T-Poxy is suitable for making fuel tanks without a special liner and is so used in the *VariEze* family.

All of the Hexcel resins of the Epolite family should be stored in a constant-temperature rather warm location (77 °F). Some builders construct a little plywood closet, insulated with cheap sheet foam from a hardware store, kept warm with a small light bulb. The ratio pump and the storage containers are kept ready for use in the warm closet. Recommended storage time for these products is one year, which is probably conservative, but we're building airplanes, here. Older material can be used in non-critical applications; use fresh material for spars, etc.

Aeropoxy

Rutan Aircraft Factory now recommends (early 1995) the use of the Aeropoxy system, manufactured by PTM&W Industries, to replace the old RAE resins and the newer Epolite. The system contains additives to improve bonding to carbon and Kevlar. Viscosity is low enough to give good wetting and handling. It contains no MDA and meets current OSHA requirements. (But take precautions anyway!) The material is clearer than Epolite and has less odor. It can be cured at elevated temperature and the finished structure is stiffer at high ambient temperatures. The product was improved in 1996; the new resin is type PR2032.

Two hardening speeds are offered. The general-purpose hardener type PH3660 gives a pot life of about 60 minutes and is mixed at a volume ratio of exactly 3 parts resin to one part resin. This combination produces a glass transition temperature, T_g, of 196 °F, so post-curing is not necessary, though remains an option for improved characteristics. With hardener PH3417, the pot life is about 20 minutes and the volume mix ratio is exactly 4 to 1. (The new system is compatible in bonding with the earlier PR2095/PH3660-2 system.) Ratio pumps in these ratios are available from the aircraft suppliers. The fast-hardening combination is not recommended for primary structures, but is to be used for repairs and additions to primary structures where high temperatures will not be encountered. (T_g is 180 °F.)

Poly Epoxy

Poly Epoxy is a product of the Poly-Fiber Aircraft Coatings Company. In appearance and handling characteristics, it is similar to Aeropoxy, described above.

It has excellent handling and wetting characteristics at lower than ideal temperatures. It was used successfully in the EAA/Aircraft Spruce Builder's Workshops in Advanced Composites in an unheated hanger in Chino, CA, in February.

The manufacturer makes a particularly strong point about the improvement in strength of Poly Epoxy with post-cure. (See below.)

Poly Epoxy is one of the most expensive matrix materials offered to the homebuilder. The company offers a companion product, Alpha Poxy, which is a much cheaper non-structural epoxy, to be used for filling and for non-critical applications.

The hardener component of Poly Epoxy is subject to spoilage from contact with water and carbon dioxide in the air, so mixing should be done by weight, where the hardener container can be covered after each use. This characteristic makes a ratio pump less attractive, unless the builder is using quite a bit of material so that long storage in the pump chamber is not required. A low cost fixed ratio pump is not offered for this ratio (40:100), so a variable ratio pump would be required.

West System Epoxy

This system also uses a base resin (#105) with a fast hardener (#205) and a slow hardener (#206). The system is primarily directed at marine applications and offers the useful properties of particularly good moisture resistance and good wetting of wood. It has relatively low viscosity and lends itself to use at a wider range of working temperatures, including cooler temperatures, than the other epoxies described above. The fast hardener gives a pot life of 10 to 15 minutes at 70 °F and partial cure at 5 to 7 hours. The slow hardener gives 30 to 40 minutes and 9 hours. Special inexpensive "minipumps" are sold for direct attachment to the West System containers to give the correct ratio of components, thus the cost of a separate ratio pump is avoided. The mixing ratio is 5 parts resin to one part of hardener, by volume. With the available short hardening time and the somewhat lower cost, this system is attractive for making molds.

"Inhibited" Resins

This family of resin materials is characterized by the fact that they are chemically complete to polymerize into a solid, but the reaction has been inhibited by a small amount of some agent. When we wish to start the hardening reaction, we neutralize the inhibitor by adding a small amount of a neutralizing agent, allowing the reaction to continue to completion.

Polyester Resins

The polyester resins are by far the lowest cost and most widely used of the several matrix materials that the builder might consider and is the material commonly sold in paint stores and auto supply houses as "fiberglass plastic." It has been almost universal in the boat industry for decades and is seen in plastic shower stalls and hot tubs. It finds wide use in low-stress applications where weight and fiber-to-matrix ratio is secondary to low production cost.

Two types of polyester resins are used: (1) orthopthalic (or "ortho") and (2) isopthalic (or "iso"). The "ortho" has better physical properties as a matrix. The "iso" is more flexible but has better adhesion qualities to fiberglass and other surfaces and has excellent resistance to petroleum fuels. Shrinkage of the finished structure is significantly greater with the polyesters than with other matrix materials.

To use, the builder pours a working quantity of polyester into a cup, adds a small measured quantity of a volatile catalyst, stirs, and applies it to the work.

The base pure polyester material is sold as *bonding resin* (or Type B) and is intended to be the base coat or an intermediate coat in a multi-layer laminate. Bond coat has the useful property of not hardening completely at the surface, leaving a thin sticky film of non-hardened plastic which forms an excellent bond with the next layer to be applied. Two explanations for this sticky film are given in the literature. In one, the polyester is hygroscopic and draws moisture from the air which inhibits the catalyst at the surface. In the second, the volatile catalyst evaporates from the surface before it can serve its function, leaving a film of uncatalyzed plastic. (Choose the theory you like.) Bond coat resin will form a hard surface if protected from the air with a layer of plastic such as Saran Wrap or polyethylene film, which is peeled off after the resin hardens.

To get a hard surface in air, a small quantity of wax is added to Type B bonding resin to make *Type A surfacing resin*. This product is sold with the wax added as Type A resin. The wax comes to the surface and seals the surface against the air, permitting it to harden. The problem is that the wax remains after hardening and must be removed by sanding or solvents or both if there is to be some other final coat, such as paint. The surface curing agent, basically the wax, is separately sold to convert bond coat material to Type A surfacing coat material, when needed.

These materials have some significant disadvantages in handling by the home builder. The catalysts are nasty, dangerous substances which must be very carefully handled every time a working batch of matrix is needed. The materials contain a high percentage of styrene (the source of the familiar "fiberglass smell" associated with new boats) and OSHA recommends that fabricators work with respirators which give protection to specification TC 23T. (Respirator cartridges which meet this requirement are generally available at paint stores.)

The author's conclusion is that these materials should be used only where the low cost is important enough to justify the poorer physical characteristics and the handling problems. Mold-making, boats, and shower stalls are good applications. Even in mold-making, shrinkage may be a problem.

Vinyl Esters

Several formulations of this chemical family are sold by Dow Chemical under the name *Derakane*. Physical properties are quite good, including properties at elevated temperatures. Derakane is used by Stoddard-Hamilton in the *Glasair* line.

The base vinyl ester resin can be stored in bulk for long periods of time. The material is "promoted" prior to use by the addition of small amounts of certain chemicals. The storage life of the "promoted" resin is of the order of a few months. The "promoted" mixture is then catalyzed immediately before final use. Derakane also contains a large amount of styrene.

Derakane has properties which make its use more advantageous to an established manufacturer like Stoddard-Hamilton (who can control the working environment) than to the individual builder. The advantages are superior physical properties, lower cost, and ability to adjust pot life to work room temperature. The disadvantages are the problems of storing promoted material, handling the catalyst, and working safely with the styrene and the dangerous additives. These problems are manageable, however, and Stoddard-Hamilton supplies "promoted" material in small batches to *Glasair* builders for use in the home shop.

Conclusion

To repeat: Unless you are very, very sure that you fully understand what you are doing, stick to the materials family chosen by the original designer of your aircraft. You will be too busy building your airplane to also undertake research in polymer materials science. If you are the original designer, use the successful system chosen by a previous designer. If you insist in pioneering in materials (and Burt Rutan did!), then get plenty of technical input from the engineering staff of the materials suppliers and be prepared to do the necessary testing.

Chapter 6
Core Materials

Introduction

In Chapter 4, we considered the theory of how sandwich structures worked and how it was that the combination of strong, thin skins and a relatively light "core" material could produce a very efficient structure. Recall that the chief purpose of the core material was to pass shear forces between the skin surfaces. The ideal core material would be one which was as light as possible while still being able to carry the distributed shear loads. These required shear loads are very much less than the tension and compression loads that must be carried by the skins because they are spread across a much larger area. In this chapter, we will discuss some specific core materials that the home builder will be likely to use or would find in a kit which offered pre-built subassemblies.

Wood

Wood makes a good core material, though it is rather heavy, compared with some other alternatives. As an aircraft core, it is nearly always used in the form of plywood. It has two considerable advantages: (1) it is cheap and (2) it is strong enough to hold its position while the surface skin is applied. We often see wood used as a "core" in small boat construction, where a builder will "glass over" a semi-finished plywood structure. The builder is usually aware that his structure has been stiffened by the glass composite skin, but may see the "glass" as primarily a surface finishing technique which keeps water from soaking the plywood. Often, he does not understand the actual function of a "core" in a structural sandwich. The builder is usually happy with the result, unless some damage (or screwholes) penetrates the glass skin and allows water to eventually soak the plywood and cause expansion and rot. The author has used "glassed plywood" many times for non-aircraft applications where a light, thin, quickly-built, reasonably weatherproof panel is desired.

Plywood cores are often used in aircraft for heavily loaded bulkheads which may carry engine mount or wing attachments.

Balsa wood, at a density of about ten pounds per cubic foot, is an excellent core material for a seriously engineered high performance application. As a core material, it is usually used with the grain perpendicular to the facings. As a natural product, it is subject to considerable variation between pieces and will require some care in selecting pieces to be used. Like other woods, is is subject to damage if water ever enters. The aircraft supply houses that serve the homebuilder offer balsa wood, but the author knows of no current homebuilt design that uses it.

Foam Core Materials

Many kinds of foams are available to the homebuilder and they can be used as cores in many different applications. They vary in density from less than two pounds per cubic foot to as much as 18 lbs/ft³. Most have names only an organic chemist could love, let alone understand. The homebuilder only needs to understand what type is required for a given purpose then follow one of the established techniques for any particular combination of foam, facing, and bonding to the facing.

Foams are particularly important materials for the home builder because they can serve *both* as cores and as forms. Used as a form, the light weight foam material is cut to some complex shape, such as a wing, then is covered with a strong skin to serve as the core in the structure.

The discussion below will briefly review the general characteristics of the several types of foam used by home builders who build from scratch with simple equipment. The well-equipped manufacturers of prepreg kits may use the same foams, but can also use others which require special tools or are not generally available.

Polystyrene Foam

This foam is very widely used as the core material for wings and control surfaces in those aircraft using Burt Rutan's original Moldless process. This includes the *VariEze*, the *LongEze*, the *Dragonfly*, and all their descendants. It is widely available in a range of sheets and blocks with a light blue tint manufactured by DuPont as "Styrofoam" at 2 lbs/ft^3. Dow Chemical makes a similar product with an orange color, but it is not available in as many block shapes as the blue foam.

Polystyrene foam has two outstanding advantages: (1) It lends itself to easy cutting and carving by means of a hot wire cutter without producing toxic gases, and (2) its shear strength and stiffness matches that of glass very well for a high performance core structure.

A possible disadvantage is that many solvents attack this material, in particular (1) styrene, a component of the polyester and vinyl ester resins and (2) gasoline and some other solvents. The first is not a problem if epoxies are used.

> *The second <u>shouldn't</u> be a problem. The author knows of a VariEze pilot who had a very narrow escape because the consequences of a very slow fuel leak into his wing that was discovered just in time. Fuel tanks must be built so that they <u>cannot</u> leak and whatever special measures and tests are needed to make this so <u>must</u> be taken. (Mainly, builders must remember that airplanes must bend a little in normal service.)*

An additional possible disadvantage of the blue polystyrene (Du Pont) is that it is quite susceptible to damage from ultraviolet radiation. In practice, this only means that the builder cannot store the material in sunlight for long periods of time. It is necessary to protect the outer skin of a composite aircraft from UV in any case, so the core will be protected. The Dow orange foam is not subject to UV damage, so might be preferred, other things being equal.

Polyurethane Foam

Many urethanes are offered to the general market for a wide range of non-aircraft applications, but only a few are appropriate for aircraft cores. Most are highly inflammable and ALL PRODUCE A DEADLY POISONOUS GAS WHEN BURNING. It is extremely dangerous to cut this material with the hot wire technique used with styrofoam. Carving and cutting may be done with ordinary tools, but the material lends itself very well to carving into complex curved parts like wheel pants or fairings. *The dust is very hazardous and a protective mask must be worn when working this material.*

Polyurethane foam is a little cheaper than styrofoam in equivalent sizes and densities and *can* be used with the cheaper polyester resins. They tolerate most solvents, including gasoline, and mechanical properties are good.

Urethane/polyester Foam

Last-A-Foam is a urethane/polyester which is less flammable than straight urethane. It is offered by Aircraft Spruce in densities from 4.5 to 18 lbs/ft^3. It has the advantage of usefulness to 180 deg. F.

Polyvinyl Chloride (PVC) Foam

PVCs are very widely used in such products as the plastic used in common water pipe and garbage cans. The only PVC foams in use in the small aircraft community are Klegecell, from Italy, and Divinycell, a Swedish company manufacturing in Texas.

This material has good mechanical properties, good solvent resistance, will work with most of the available matrix materials, and will tolerate the 250 deg F in the curing oven required by prepreg manufacture. It can be formed in compound shapes at 200 ºF in a vacuum mold.

The smaller *Lancair* designs use Divinycell in the tail surface skins, where the core spacing between facings is small.

Aircraft Spruce offers Divinycell at densities of 3, 6, and 15.6 lbs/ft^3.

"Rohacell"

This is a very high performance material at 1.9 lb/ft^3 with very fine cell size and excellent uniformity. It may be used in curing ovens up to 350 ºF and in service at up to 250 ºF. It may be used with nearly all resins and will resist solvents. It is, of course, more expensive. The primary disadvantage is that the tools necessary for practical fabrication into useable sandwich structures are not available to homebuilders. This foam is excellent in a fully-tooled industrial application where heated closed cavity molds can be used.

Honeycomb Cores

A "honeycomb" core is an array of hollow columns made of sheet material which is used to separate the two facings of a sandwich structure. The similarity to the honeycombs made by bees is so obvious that no other name could suggest itself. The sheet material used to form the hollow columns may be paper, woven fabric or metal.

To manufacture the core material, layers of the raw sheet material are bonded to each other along lines at regular intervals. The bonded stack of layers is then sliced to the desired core thickness. The sliced stacks are then pulled apart to expand the honeycomb core pattern. Like the honeycomb produced by bees, the columns produced by this method are very nearly hexagonal. The hexagonal pattern is produced by what amounts to long lines of corrugated ribbons which touch on the faces of the corrugations.

Bearing in mind our brief review of the theory of structures, we see that the honeycomb core is a mostly-hollow network of little ribbons, loaded edgewise. The hollow network is even less dense than a corresponding foam material which could have been used as a core, but the little ribbons can stand higher stress than the foam, even though the foam is bonded over its entire surface. The function of the core to distribute shear stresses between the faces is the same, so the thin ribbon edges must be very well bonded to the facings. This bonding function requires excellent tooling and very good control which is not generally available to the home builder.

The skilled professional designer can choose to vary the pattern of the honeycomb columns to achieve different strengths in different directions. Very high performance can be achieved. One core design uses foam to fill the the honeycomb cells to stabilize the columns against buckling, but most are hollow.

Honeycomb-cored sandwich structures are widely used in the commercial aerospace industry, especially where very high performance is required. For the most demanding aerospace applications, exotic materials can be used, including titanium, carbon fiber, and Kevlar. Cost and required tooling makes these materials impractical for the homebuilt aircraft application.

Honeycomb cores made of brown Kraft paper are widely used in the manufacture of residential lightweight "hollow" doors, as sold at the neighborhood home supply store. The paper and the thin wood facing is sensitive to moisture problems, so these doors are only useful for residential interior applications. They are, however, light, stiff, relatively stable, and cheap. (And they make great temporary work tables in the shop!)

Nomex paper, made by DuPont, is fabricated into honeycomb core material and is used by the aerospace industry for a wide variety of applications in cabin interiors. Aircraft kit builders use it in primary structures to make wing skins, bulkheads, and fuselage sections.

All honeycomb structures are made by placing sheets of adhesive material between the expanded core sheet and the facings. The resulting sandwich is then pressed into final position in a tool and then heated to make the adhesive bond.

Clearly, the proper design and fabrication of honeycomb core structures is a job for professional engineers. A kit manufacturer will have the necessary tooling in place to make reliable large honeycomb core structures, like wing skins. This fact makes honeycomb core technology available to the kit builder where he would otherwise have to use a foam core, if building from scratch.

Chapter 7
Joining Structures

Adhesive Joints

The joining problem arises when we must connect large separate structures like upper and lower fuselage halves or spar assemblies into a wing, where there may be small gaps and misalignments.

We use adhesive materials to join the large structural subassemblies that are assembled to make a composite aircraft. One particular advantage of composite materials is that we can safely obtain really good long-term performance from bonded composite joints, using simple techniques available to the home builder.

In contrast, we see that the mainstream aircraft industry has successfully used bonded metal structures since WWII, but has required complex chemical treatments and extraordinary cleanliness to achieve reliability. With aluminum, this is because the chemically active surface reacts with air to form a soft oxide which is far weaker than the bulk metal. Reliable bonding techniques require that this oxide be removed and kept off during the bonding operation. It can be done in a homebuilt environment, but the work is hard to inspect and the consequences of spectacular failure are most unattractive.

Basic Requirements of an Adhesive

Remembering the "puzzle" at the beginning of Chapter 2, consider what is required to make an overlapping lap joint between three smooth rectangular blocks. We'll use a matched pair of adhesive joints so that the system will be perfectly balanced - no question of anything other than pure shear forces on the joints. See the sketch following.

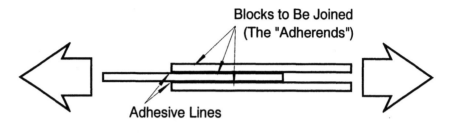

Adhesive Test

Now suppose that the blocks to be joined were - say - automobile tire tread rubber and that the adhesive is "really strong stuff" that has the physical characteristics of steel. Stress this specimen in a test machine. What happens?

The joints break progressively along the adhesive lines at the rubber face. See that as the stress is applied, the rubber at the edge of the joint stretches (strains). The stiff "steel" glue does not produce much strain, so the attached rubber (which must have the same strain) cannot develop much reaction force. The rubber at a distance from the joint indeed stretches. The result is an extreme stress concentration which works like a zipper to peel the rubber right out of the joint. This is called *adhesive failure*, because the glue and the "adherend" separated.

Now reverse the properties of the two materials. Let the blocks be steel and let the adhesive have the properties of automobile tread. Apply a stress. The steel strains (but not much). At the "glue" joint, the "rubber" must strain sufficiently to generate a shear stress which produces a force equal to the total load. Given that the area of the glued joint is large relative to the cross-sectional area of the steel, quite a large total shear force can be transmitted, perhaps more than the steel can carry. In fact, a rubber-like material might be quite a practical adhesive for this case. Assuming that the total strength of the adhesive is actually less than the strength of the steel, if the test force were increased to failure we would see *cohesive failure,* or failure within the material of the glue joint. With sufficient bonded area, the steel part of the specimen would fail first.

Joining Wood

We don't see much wood used in highly stressed structures in composite designs, but we do see it used as a sandwich core material and as a support for molds. Some builders still prefer it to other materials for primary structures in spite of the previously discussed disadvantages. Wood has been used for aircraft since the Wright brothers and a great amount of published standard practice has evolved. Wood was used in WW II when aluminum was in very short supply and appropriate wood production facilities existed. Modern glues are available which are far more environmentally stable than the woods they are to join. (It will be assumed here that a builder who intends to use wood as primary structure in her aircraft will go to published sources for wood adhesive techniques!)

The matrix resins used in layups can be good adhesives, but the low viscosity that allows them to penetrate a layup also makes them squeeze out of a joint if much pressure is used to force alignment, as it often is. A little bit of micro (glass beads) can be added as a filler to set a minimum thickness for the glue line. Also, the moisture inevitably present in wood may reduce the properties of the epoxy.

While it is convenient for the composite homebuilder who is using his matrix material for practically everything in the shop to use it for wood glue, too, if the application is critical it may be better to go to formulations intended for marine use. The West System materials are primarily sold as water-resistant marine adhesives, but are offered by the aircraft supply houses.

Joining Formed Composite Structures

The butt joint is the simplest way to join two composite structures. The two structures are brought together, held accurately in position, and essentially "repaired" to make a new continuous structure, as shown below.

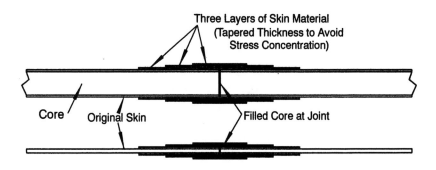

Simple Butt Joint - Core and Single Skin Cases

To make a butt joint for a sandwich structure, we essentially replace the missing or damaged core material and add new skin material sufficient to carry the required stress across the joint. As shown, the added skin material should be tapered to avoid stress concentration near the joint. The single-skin case is similar, except that there is no core to fill. (We assume that a single skin structure is not required to carry bending stress, but will only receive stress in the plane of the surface.)

The simple butt joint has three important disadvantages: (1) The sketch exaggerates the thickness of the added skin, but this method usually makes a noticeable bulge in the finished surface, and (2) it requires full access to both sides of the joint (which might be difficult in a fuselage tail cone), and (3) requires that the parts be located very accurately.

Where it is practical to cut away some core material before laying up the skin, the core material may be recessed near the joint and the joining doubler may be placed in the recess, as shown below.

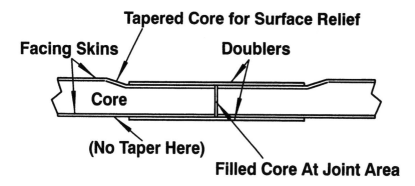

Simple Butt Joint With Relieved Surface

The disadvantage of requiring full access to both sides of the joint remains: essentially, the builder must do a complete layup on both sides to make the joint, which is often impractical for a long structure.

The tapered core lap joint shown below mostly avoids these disadvantages and is typical of the method generally provided by composite kit manufacturers.

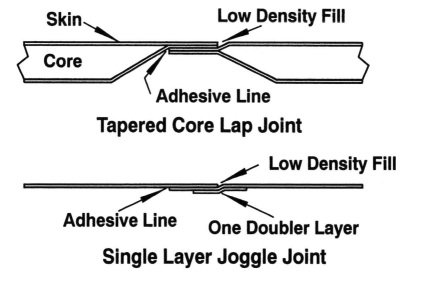

The previous illustration shows the case where an additional doubler skin layer has been added to each of the joined sections to strengthen the joint where the core is tapered. See that the technique provides a very large adhesive area to make a very secure joint. To assemble, the joint may be clamped or held with a line of "pop" rivets until cured. (After cure, the pop rivets can be drilled out and the holes filled, with negligible loss of strength.) Alternatively, this joint can be made removable by using bolts to hold the structures together. (If bolts are used, care must be taken that adequate bushings and washers are provided to avoid crushing the lap area at the bolt holes. See the section following on bushings.)

The precision molded sandwich structures offered by the kit manufacturers always provide for bonding areas to join the parts. Often, the edges of a structure are made with the core tapered to zero thickness and the inner and outer skins brought together to form a single smooth surface area for bonding. If necessary, the designer may specify additional reinforcing layers at the transition between the sandwich skin and the joint. The corresponding mating structure will then have the joint joggle in the opposite direction so that the two parts fit together to give a smooth outer surface. (Here, the similarity to the toy plastic snap-together models is quite apparent.) Where it is not practical to apply significant bonding pressure, adhesives such as Hysol 9330.3 or 3M "Scotchweld" 2216 A/B Gray are used, held in place by Clecos or by pop rivets.

Alternatively, depending partly on the accessibility of the work area, added doublers of compatible composite material may be placed over the area to be joined. Here, the designer of a molded part would allow for the added thickness of the doubler skin. Doublers work with either single skin or sandwich core designs.

Bushings and Attachments

Bushings are structures used to pass loads from a small area of high load concentration, like a bolt, to a larger area with less load concentration, like a bulkhead. Suppose we had the problem of bolting an engine mount fitting to a fuselage bulkhead. Typically, a bulkhead is a light, thin-walled, flat structure that is bonded to a large circumference of fuselage surface. The engine mount bolt is usually of medium-hard steel of a size chosen by the designer as safely adequate for the relatively large forces to be carried. We cannot simply drill a bolt-sized hole in the thin-skinned bulkhead, insert the steel bolt, and go. The concentration of stress at the side of the bolt would cause local failure of the bulkhead. In addition, some kind of sleeve is necessary to carry the tensile force of the tightened nut and avoid crushing the core.

One solution would be to provide an oversize steel bolt, one whose radius is sufficiently large to reduce the stress on the thin bulkhead skins to an acceptable level. This would usually require a grossly oversize and thus an over-heavy bolt.

A better solution would be to provide sufficient added skin thickness at the bolt to reduce the skin stress at the bolt to an acceptable value. The bolt load must then be dispersed to the surrounding thin bulkhead skins. This can be done by covering a larger area of bulkhead skin with additional material. To avoid excessive stress concentration, the skin thickness should be tapered to match the skin of the bulkhead. See the sketch below.

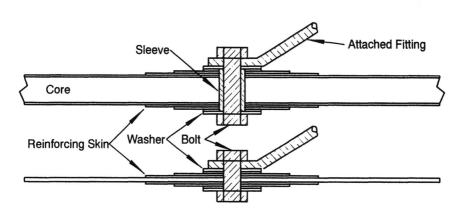

Simple Bolt Attachment - Core and Single Skin Cases

A variation of the first case shown in the sketch (used in the *Dragonfly*) is to provide a larger bushing of a lighter material, like an aluminum or phenolic disk, to increase the area of the interface between the bulkhead skin and the high stress area, allowing the use of a smaller bolt.

One easy way to make a bushing for a sandwich structure (as used in the *KIS*) is shown below. Here, the skin and core material is removed from one side, leaving one skin intact. A piece of phenolic laminate is cut and bonded against the intact skin. Damaged core material is replaced with "micro" filler and new skin material is lap bonded over the assembly. After hardening, the assembly is drilled to accept the bolt. This method is appropriate for moderate loads. See that the phenolic insert takes the direct load of the bolt and carries the load to the skins near the edge of the phenolic insert.

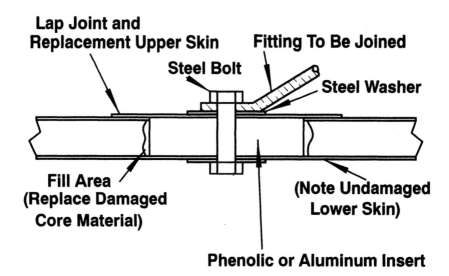

Bolt Joint in Sandwich Core Structure

Bonding Continuing Layups

The Secondary Bond

The ideal situation in construction of a composite aircraft would be to make every separate part as one giant construction where, for example, an entire wing would be laid up and completed in a single work session. In reality, it will frequently be necessary to make a layup over a previously-cured surface where we require that the surfaces be permanently bonded *without having reinforcing fibers pass through the bond.* This is a *secondary bond.*

The secondary bond will be designed so that the non-reinforced joint is loaded almost entirely in shear and is provided with generous area to carry the required load. Usually the adhesive is the same resin used in making the layups to be bonded. The bond, then, is essentially a continuation of the layup, except that there are no fibers across the joint.

When it is necessary to join large parts which are already cured, structural adhesives are used which have a lower stiffness than most layup resins, so that shear loads are carried to a large area. (Recall the discussion on page 7-2.)

Examples of techniques for making accurate large-area joggle joints for secondary bonds will be discussed in Chapters 12-14.

Chapter 8
Solid Core Structures

Flat Sandwich Structures

A practical structural sandwich consists of two thin, strong surfaces separated by a relatively thick core material which is intimately bonded to the surfaces. We discussed the characteristics of simple sandwich forms in Chapter 4. In aircraft construction, the surfaces are usually strong composite or aluminum sheets and the core may be one of the several very light materials we discussed in Chapter 6. Flat sandwiches are widely used in commercial aircraft as bulkheads, floors, panels, and as components of larger structures. In home building, the builder usually laminates the sandwich himself as needed, though prefabricated flat sandwich stock is available to kit manufacturers. (The aircraft supply houses that normally supply the individual home builder do not carry prefabricated sandwich stock.)

Burt Rutan's Composite Sandwich

We have discussed an ideal sandwich structure in terms of two flat plates separated by a uniformly flat core. Alternatively, the core may be shaped as a surfboard or some other complex form. In fact, the surfboard designers pioneered the use of a strong composite skin over a shaped lightweight core. The modern surfboard is very light, very strong, and survives in a corrosive environment.

In 1975, Burt Rutan extended the shaped-core techniques to aircraft with his hugely successful VariEze. He characterized the method as *Moldless Composite Sandwich Homebuilt Aircraft Construction*. In fact, the shaped core is the internal "mold" that determines the shape of the surface, so the builder has the task of finishing the external surface of the final structure. Rutan's tech-niques were picked up by other designers and continued in a number of very successful aircraft.

Designs of the Rutan family make use of two very effective forming processes: (1) quick, accurate cutting of large foam core parts to a final shape, and (2) construction of a curved sandwich shell which consists of inner and outer skins which are separated by a core of foam. Most of a "Rutan-type" fuselage may consist of this formed shell. The shell gets its stiffness from its sandwich form with thin light skins in contrast with the typical fiberglass boat, which simply adds more material to a single thicker skin until it is "strong enough" to resist buckling. In both forming methods, the outside skin carries an air load and must be finished to the final smooth aerodynamic surface.

The Solid Core Airfoil

The Rutan-type designs make use of a foam core which is accu-rately cut to an airfoil shape then covered with a composite skin to make the completed part. This method is used to make entire air-foils, such as control surfaces, and no other internal structure (such as a conventional "spar") is used for these smaller parts. The part may be considered to be a wing-shaped sandwich beam with con-stant skin thickness, where the core tapers to zero thickness at the edge, as shown below.

Uniform Skin Thickness

Core

Solid-core Airfoil

The Rutan-type designs are probably unique in aircraft construction in providing a completely distributed structure to carry distributed loads. The shear stresses are distributed over a very large volume compared to, say, a World War I design which ran most of the load from the fabric covering into one or two spars through a complex network of tiny parts.

For a wing structure of a size typical of small aircraft, the skin may be only a small fraction of an inch thick while the continuous core is several inches thick. It is usually not practical to vary the skin thickness significantly. The middle part of the airfoil section, where the core is thick, is an efficient thick sandwich structure. Nearer the trailing edge, the *core* thins, but the *skin thickness* is constant, so this part of the structure contributes less to the overall stiffness.

Where necessary to carry a larger load, additional skin material can be used at the widest, higher stressed, regions for a structure with uniform core material, as shown below.

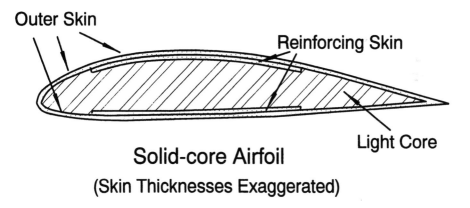

Outer Skin

Reinforcing Skin

Light Core

Solid-core Airfoil
(Skin Thicknesses Exaggerated)

This simple solid core airfoil is used in several designs for small control surfaces. No current solid-core aircraft design makes use of large sections of a wing which do not use some internal structure in addition to the foam core.

Suppose we simplify the design problem by ignoring concentrated loads like landing gear attachments and the need for fuel tank cut-outs, and consider only the comparatively simple structure of an outboard wing section. This section must carry the distributed flight loads back to a distributed attachment at the inboard section of the wing. The flight loads increase from nearly zero at the wing tip to a maximum at the attachment to the inboard section. It is theoretically possible to provide the optimum amount of reinforcing fibers in the skin to carry the widely distributed tension and compression loads, as suggested in the sketch above. In practice, we can add skin material almost without limit to meet the tension and compression requirements, but we can't adjust the characteristics of the foam, if a single foam type is used. There will be a maximum load at which the lightweight core fails before the skins fail.

One way to avoid this failure would be to go to a stronger foam in the region of high shear. Candidates would be Divinycell PVC at 3 lb/ft^3 or Urethane/polyester at 4.5 lb/ft^3. It would be interesting to build test samples of "sparless" wing sections of this type and to perform actual tests to destruction to find the limits. This scheme would preserve much of the simplicity of construction of the solid core wing, while adding weight only where it will carry a useful load. See below.

Outer Skin   Stronger Core
Reinforcing Skin
Light Core   Light Core

Solid-core Airfoil
(Skin Thicknesses Exaggerated)

The Embedded Spar

Actual aircraft designs, such as the *VariEze, LongEze,* and *Dragonfly* do use a spar, as shown below. The front of the wing section is cut from a single piece of foam, leaving an indentation on the top and bottom, as shown. A composite shear web is laid up on the back of the foam section. The indentations hold several layers of fiber to make a spar cap. The "C-shaped" assembled spar is then what structural engineers call a "channel section".

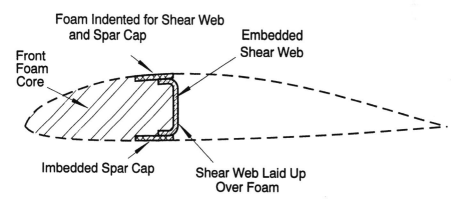

Front of Solid Core Airfoil With Embedded Spar

The rear of the wing core is then cut from a second block of foam and attached as shown. See that a notch is provided to hold several layers of fiber to make a spar cap. This notch is filled to the level of the wing skin. The entire assembly is then covered with the final skin to complete the wing, which now includes a strong composite shear web completely immersed in lightweight foam. Shear loads are transferred between the spar cap and the flat area of the shear web. See the sketch on the next page.

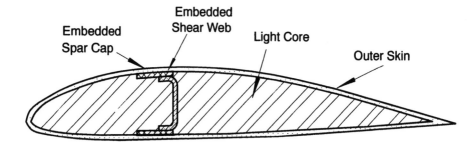

Solid Core Airfoil With Embedded Spar

(Skin Thicknesses Greatly Exaggerated)

Recalling the introductory puzzle in Chapter 3, we see that the "soft" foam that surrounds the "stiff" spar essentially receives very little load - the spar gets nearly all. The nearby foam serves mainly to stabilize the thin shear web against buckling. Also, the mechanism for passing shear loads between the top and bottom skins has changed. Tension and compression loads are shared between the wide skins and the relatively narrow spar cap. The "soft" foam next to the spar will not receive enough strain to generate an effective reaction force. The foam will only be effective in passing shear loads at a sufficient distance from the spar to receive strain from the skins. Much, if not most, of the total shear must be carried by the imbedded shear web. This is a highly redundant structure with mixed material stiffnesses and accurate analysis of the forces in the system is a challenge for the very best structural engineers.

This method for constructing a wing has several outstanding advantages, which is probably why it has come to be so widely used. The principal advantage is that it is relatively easy for the amateur builder to build a strong, safe wing. There aren't too many separate parts or operations and the cut foam is its own form. A purist might argue that the method might be optimized by reducing skin thickness and using more imbedded spars with less material per spar. It would be hard to prove that the added benefit would be worth the added work, however. See the sketch on the next page.

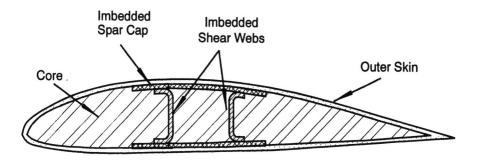

Solid Core Wing With Two Embedded Shear Webs

(Skin thicknesses greatly exaggerated)

Another design variation should be noted. Some designs, including the *Dragonfly*, use stiff carbon fiber for spar cap reinforcement. By the previous argument, see that the glass skin (lower-E) in the vicinity of the spar cap is no longer effective in carrying a significant share of the bending loads and the design now depends on the high-E carbon for its performance, permitting the designer to use a thinner skin.

Any solid core structure will add weight to the finished structure in proportion to the volume of core material used. Whether this is a good trade-off over going to a hollow structure depends on the specific case and what the practical alternatives are for the materials available. If a wing design is being considered, the core weight will increase as the cube of the wing size and, even with foam at 2 lbs/ft^3, there will be a point at which a hollow structure will be lighter for the same design strength.

Chapter 9
Basic Moldless Tools
and Techniques

This chapter will discuss the basic techniques used by the home builder in current composite aircraft designs. These techniques require only very simple hand tools. Plans for established and tested designs will have more detailed instructions for each operation and, as always, those instructions should be followed. This chapter will be a summary of the methods originally described by Burt Rutan (and modified by his successors), and is intended only to provide a general understanding of the principles. The details of the design of strong points like wing attachments and bulkhead fittings will be specifically engineered for each aircraft and will not be discussed here.

If a reader wishes to undertake the necessary engineering work to produce a new design using these techniques, she would find it useful to study the details of how preceding successful designs were made. Rutan's organization no longer sells *LongEze* plans, but plans for the *Dragonfly* are available from Viking Aircraft. The *Dragonfly* is a direct descendant of the *LongEze* and the detailed instructions are fully as good as Rutan's.

Rutan's Original Construction Manual

It is very hard to improve on Burt Rutan's original *Moldless Composite Homebuilt Sandwich Aircraft Construction*, introduced in 1978 as a part of the plans of the original VariEze, and revised for the LongEze in 1980. The book was later sold separately, either alone or as part of an introductory kit to expose the prospective builder to these then-radical techniques. It is still available through the aircraft supply houses. The book is quite short, is extremely well illustrated, and clearly describes the methods that convinced thousands that they could successfully build a good airplane. All of the basic methods for handling fiberglass, mixing epoxy, cutting foam core material to shape, applying epoxy, and making a layup over foam are carefully described, including a practice exercise.

Many things have improved since 1978-80. Rutan's original matrix resins have been replaced with better materials (Chapter 5) and today's builders will generally agree that measuring epoxy components with a balance scale just isn't worth it compared to using a ratio pump. Also, most of today's builders control the hot wire foam cutter with a simple combination of a household light dimmer and a transformer. (The inexpensive hot wire controller is now sold as a kit by the suppliers.)

The author strongly recommends that Rutan's pioneering book be obtained by any reader who is seriously considering a composite project and wants to get right down to the bottom of the grubby details. Better, the cautious reader should get the introductory kit and Rutan's book together for about $50. Even if the reader never builds an airplane, but only uses the methods to make a weather-proof storm door, he will be the wiser for the experience and will likely add composite techniques to his inventory of useful skills.

The Bench

You will need a long, stiff, table many times during the course of construction. The preferred dimensions depend on the design to be built and the nature of the workspace available. Rutan recommended a table 12 ft. long, 3 to 4 ft. wide, and 35 to 39 inches high. Experience of later builders suggests that longer is better and that the table should be narrower to allow access to the work without having to lean over the edge. It should also be a little lower than Rutan's height. *Dragonfly* plans recommend a table 20-22 ft. long, only 16 inches wide, and 32 inches high.

Whatever the builder's choice, the table should be a long, stiff, hollow box made of unfinished 5/8 inch particle board attached to straight side pieces 4 to 6 inches wide. The table cannot be made accurate enough to serve as a primary reference surface or a jig, but should be as close to a true flat as is reasonably convenient. The rough surface of the unfinished particle board is very handy for attaching temporary braces and jigs with Bondo. After use, the fixture is knocked off of the particle board surface. Fragments of particle board will come off with the fixture, but this is not a problem. We don't mind a little low spot; we just don't want a bump. Most work can be done on the narrow table. When a much wider table is required, light "hollow" doors, cheaply bought at the local home supply store, can be placed over the bench, as needed, and stacked against the wall for storage.

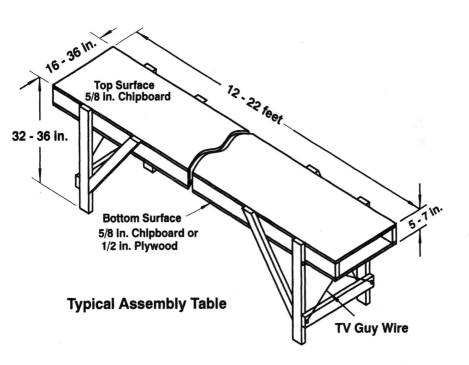

Typical Assembly Table

Small Tools and Supplies

Following is a partial list of the essential tools that will be referred to in the following discussion of techniques. The aircraft supply catalogs show these and a great many others which the individual may decide are nice to have.

Mixing Resin

Ratio pump or scale.

Wax-free flat-bottomed mixing cups: 3 oz, 8 oz and 16 oz. (Mostly use smaller cups)

Medical tongue depressors: In 500s, from any medical supply house

Storage cabinet: heated with light bulb

Cutting Fiber and Foam

Heavy duty fabric shears: with offset handle.

Rotary "pizza" cutter: large and small, with replacement wheels.

Utility knife: with extra blades.

Single edge razor blades.

Hacksaw: fine tooth blades.

Band saw: (Very handy, but not essential. Oscillating saws leave rough edges, orbital saws are better.)

Manipulating Wet Resin

Squeegees: Many will be used. Rubber and plastic squeegees are sold at auto body supply shops. Small ones can be cut from cardboard or plastic coffee can lids.

Layup rollers: Similar to a paint roller. Nice, but not essential

Paint brushes: disposable: 1 inch, 1 1/2 inch, 2 inch. Some builders trim the bristles short.

Sanding

Sandpaper: A wide range will be used as called for by the individual design

Rubber sanding block: Use frequently for standard sheet sandpaper

Utility cloth: In rolls. Use with long sanding sticks for leveling surfaces

Portable electric belt sander: (handy, but not essential)

Stationary electric belt sander: (very handy and nearly essential!)

Protection

Take precautions! It is essential to protect against dusts, chemicals, and flying objects. The supplier catalogs show an extensive range of protection items.

Respirator: best quality: protects against both dusts and fumes. Opinion is divided on whether protection is required for most resin materials. Generally, most resins do not require a respirator if working in a well-ventilated area. Be aware that the danger is mainly from the hardener component. Older epoxy formulations may contain MDA, which requires extra caution. Follow the manufacturer's recommendation.

Gloves: Disposable medical examination gloves are used by most builders. Cheap, not too confining, but occasionally break. Heavy duty butyl or latex gloves offer reliable protection, especially when working with solvents, but don't permit fine detail work.

Skin protection cream: Several brands are offered. Rub into skin before exposure. Wash with water after exposure.

NOTE: Use EITHER gloves or skin cream, never both.

Safety glasses.

Cleanup

Skin cleaner: Several brands, sold for removal of epoxy, paints, and greases

Solvents for tool cleanup: Acetone is widely recommended, but is nasty, toxic, highly inflammable stuff. Works fast. Expensive. Dangerous.

Vinegar: Almost as effective as acetone. Use the cheapest grocery store white vinegar. Non-inflammable. Water rinse.

> *It is not widely known in the builder community that vinegar can be used as a cleanup. Experiment with the specific resin recommended by the designer of your aircraft.*

NOTE: NEVER use a solvent, acetone or vinegar, to clean resin from skin. The solvent carries the chemicals into the skin. Use skin cleaner and soap and water.

Storing, Handling, and Cutting Fiberglass

Fiberglass cloth comes from the supplier in rolls, wrapped in paper. Recalling the importance of the "finish" to obtaining a good bond with the matrix (Chapter 5), it is important to keep the material clean and to use fresh material wherever flight loads are involved. Unfortunately, it is very hard to control dust, loose fibers, or drips of resin in a working composite shop, garage area, or hangar. It is reasonable to be careful about keeping fiberglass rolls thoroughly covered unless the material is actually being removed for cutting. Some builders use plastic garden trash bags pulled over each end of the roll and sealed with rubber bands. Others build a little covered storage cabinet from scrap plywood.

Ideally, a builder would have a large clean room used just for storing, measuring, and cutting glass cloth and separate from the actual assembly area. Composite designs frequently use long strips of material which are cut at 45 degrees to the weave of the cloth, so it takes quite a bit of precious floor area to allow the roll of cloth to be laid out for cutting. In the ideal case, a freshly-cut piece of glass could be cut in comfort on a large-sized table and be moved directly to the layup area with a minimum of exposure to contamination. (The use of stitch-bonded fabric instead of cloth, where appropriate, considerably eases cutting and handling.)

Builders in less than ideal places will find it more practical to finish all the glass-handling work before beginning each layup job.

Cutting Fibers

In preparation for glass-cutting, the builder would make an effort at cleaning up the shop, moving things around to make a cutting area. One way to make a large cutting table is to place several two or three "hollow" doors, together over folding saw horses, the main shop bench, or a temporary frame. The cloth can be rolled out on this clean surface and marked for cutting with a felt-tipped pen. It is important to use a large enough area to support the cloth because it is easy to distort the weave and lose the desired fiber direction. As each piece of cloth is cut, as called for by the designer, it is rolled up without distorting the cloth, marked for identification and put in a clean container, bearing in mind that the next hand to touch the piece may be gloved and messy with liquid resin. Don't label with masking tape because it distorts the cloth when removed.

Actual cutting can be done easily with ordinary scissors. Dressmaker's scissors (with an offset) make it much easier to follow a line without lifting the cloth from the table too much. Glass dulls scissors quickly, so either sharpen on the job or expect to buy more scissors as needed. A dressmaker's carbide wheel cutter ("pizza cutter") is very handy for long cuts or cuts which require changes of direction. To use the "pizza cutter", it is necessary to place a sheet plastic support under the cut line. The clear plastic sheet sold in hardware stores for "burglar-proof" windows works very well.

After the measuring and cutting operation is complete, and all the clean layup pieces are ready, the cloth roll is returned to safe storage, and the cutting table can be put away.

Peel Ply: Bond Preparation

Peel ply is lightweight (2.7 oz.) polyester or nylon fabric which has the remarkably useful property of NOT adhering structurally to a layup. It is used in a layup exactly as if it were a layer of structural material. The resin is squeegeed out through the peel ply and the surface is handled, inspected, and trimmed using the same techniques as in the rest of the layup. After the resin has cured, however, the layer can be removed (peeled away). The exposed surface is clean and textured and suitable for making a reliable bond to another resin surface. While it is true that a cured surface can be prepared for bonding by sanding, the use of peel ply in an attachment area is a great saving in work and avoids any chance of breaking reinforcing fibers by sanding. The builder can leave peel ply in place indefinitely while working on other parts of his aircraft, then remove the peel ply only when he is ready to continue and is ready to expose the fresh, clean surface for bonding.

The material is very widely used in the composite industry. It is sold as tape and yardage by the aircraft supply houses. Peel ply is not available in the crowfoot or satin weaves that permit other fibers to conform to double-curved surfaces, so narrow overlapping strips must be used for this situation.

Peel ply is also used to cover the exposed ends of layup plies to make a tapered edge without leaving bumps and voids or excess resin. See the sketch on the next page.

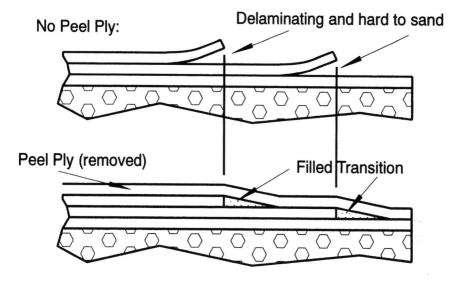

Peel Ply Transition

Some sources recommend the cheap "100% polyester" lining material sold at fabric shops for curtain lining, but use of this material is risky. Some of this material has been silicone treated and will leave a residue of silicone when removed, producing an unreliable bond. Also, some liner materials contain a small quantity of cotton which will become a part of the layup. It is safest to obtain peel ply from regular industry sources.

Peel ply should be used where a layup must be interrupted and where a bond is required when the layup is continued. An example is the need to do one surface of a wing at one working session, lay up the other surface after the first surface has cured, and make a reliable bond between the two layers. Peel ply is laid up over the areas to be joined in the same way as a structural lamination, then is removed when the next layup is made.

When peel ply is stripped from a surface, tiny fragments of peel ply material may be left on the bonding area. Most designers ignore this small contaminant, but the perfectionist builder may wish to clean up the bonding area by buffing it with a light fiber scrubbing pad like Scotchbrite. Other designers specify that the joint area be wiped with acetone.

Storing and Handling Base Resin

Nearly all composite aircraft designs that use wet layup techniques are built with a single resin type, chosen from a "system" as described in Chapter 5. If using two-part epoxies, the builder will typically have a single ratio pump set up where it is convenient to dispense small quantities of measured material as needed. In the long run, a significant amount of resin is saved with a ratio pump because it is consistently accurate for small quantities and the builder is not too concerned over mixing enough for any task. He can always conveniently mix another small batch. It is worth the trouble to keep the ratio pump and a reserve supply of epoxy in a little cabinet kept warm (85°F) with a light bulb. (The warm epoxy mixes faster and penetrates a layup easier.)

Most builders prefer the ratio pump where they are using a single resin system for the entire project. Where more than one resin system is to be used or where one of the resin components does not store well in the presence of air, an electronic scale with a direct digital readout is very useful. The Pelouze scale ($65) is battery powered and has an automatic zeroing feature which is handy when sticky material has dripped upon the measuring table. Keep a simple chart near the scale which shows how much of component "B" should be added if the builder has already poured an odd quantity of component "A" into the cup. This avoids the errors of math under pressure.

Non-structural Fillers

"Micro" - The Foam Interface

In Chapter 4, we saw why it was important that the skins of a sandwich structure be firmly attached to the core and why the integrity of the core must be unbroken if the final structure is to reach full strength. Foam core material comes in blocks that are often smaller than the final structure, so it is necessary to join pieces of core material. We need a repair and joining medium that can flow and adjust to irregularities in the joint, then harden to make a continuous foam-like structure.

"Micro" is the builder's term to describe a mixture of the matrix resin and tiny hollow glass bubbles, or "microspheres". In bulk, the dry bubble material appears as snow-white dust and is astonishingly light. ("Micro" is also used, somewhat inconsistently, to refer to the dry bulk glass bubbles as obtained from the aircraft supplier.) When the dry bubbles are mixed with the base resin, they displace many times their weight in heavy resin and make the mixture as viscous as desired. The combination of resin and bubbles makes a very convenient plastic foam which can be used as a light weight filler and adhesive for foam.

"Thin micro", or "slurry", is a mixture which has been prepared to about the viscosity of thin gravy - thin enough to flow easily, but viscous enough so that a thin coat will not run down from a vertical surface. Slurry is a mixture of roughly equal volumes of resin and dry bubbles. This mixture is runny enough to be applied with a brush. (A good guide for slurry is to see if the mixture will just run off of the stirring stick under its own weight.)

Thin micro is also used to fill the pores of the surface of polystyrene foam core before laying up the skin. It saves weight by avoiding the excessive absorption of liquid resin when the skin is laid up. Urethane foam does not require this filling operation.

"Wet micro" is about two to four volumes of bubbles mixed with one volume of base resin. It has about the consistency of honey and a thick coat will slowly run down a vertical surface.

"Dry micro" uses about five volumes of bubbles with one volume of resin. Dry micro has the property of standing up like whipped cream or pastry meringue and does not flow. It is just wet enough to have a sticky surface. Dry micro is used as a general purpose foam filler in many applications.

The art of mixing micro to suit the need is very quickly learned by direct experience. If a batch is too wet or too dry, it can be adjusted by adding the needed material and mixing again.

Special note must be taken of the fact that micro is a low strength filler only and is used only in contact with foam or as a repair or substitute for foam. Micro must never be allowed as a contaminant wherever there is to be a bond between load-carrying fiber layers. If micro should get into a fiber lamination area, it must be carefully and completely removed.

Burt Rutan's *LongEze* used only micro bubbles in the base resin for all foam filling and foam adhesion applications. Since Rutan, several other special light weight fillers with the fill material pre-mixed have become available and are shown in the supplier catalogs.

Structural Fillers

Flox

Cellulose fiber and the resins typically used in composite aircraft construction make a very good two-phase material. (See Chapter 3). Very fine (and thus strong) cellulose fiber is sold by the aircraft supply houses as "flocked cotton" and informally called "flox" by builders. When "flox" is mixed with resin so that the mixture is viscous enough to stand up like sticky meringue it makes an extremely useful structural filler. It is often used to reinforce a sharp corner and provide a moderate turning radius under a layer of reinforcing cloth. (Builders use the term "flox" to refer to both the pure flocked cotton fibers and to the mixture.) Note the differences between flox and micro: Flox is a load-carrying true composite structural material and is relatively heavy; micro is a filler only assumed to have strength comparable to foam and is mixed as light as is practical for the application. Some designs call for flox mixtures with some micro added to save weight where intermediate strength is required.

Milled Glass

Milled glass is essentially "glass flox". It consists of glass fibers which have been cut to very short lengths so that the result appears as a kind of powder. When milled glass is used instead of cotton flox as a resin reinforcement, the resulting composite is stronger than flox, but heavier. Fiber orientation, of course, is basically random, though brushing out tends to align the fibers. Milled glass is a rather nasty irritant with a tendency to penetrate the skin. Few designs call for milled glass. Wherever there might be a load concentration high enough to suggest that a very strong filler is required, the designer usually provides some more-efficient external reinforcement.

Joining Foam Blocks

To join two sections of foam core, the joining surfaces are first spread smoothly with a light coat of slurry. A "glob" of wet or dry micro is placed near the middle of the joint. The pieces are then brought together to squeeze the glob of dry micro, driving the air toward the outside and spreading the micro through the joint volume. The pieces should be slid around a little to drive the excess micro out of the joint and assure that there are no voids in the assembled core. The pieces are then held accurately in position while the matrix hardens. It is important to remove all excess micro that has flowed out of the joint, because it is much more difficult to remove after it has hardened.

"Prepreg" Strips

It is frequently necessary to apply strips of reinforcing fiber to small areas, such as wing rib attachments. In principle, a builder could cut a small strip of dry cloth, put it in place, and wet it out. In practice, this is nearly impossible, especially in the usual case where the cloth is cut on the 45 degree bias.

The simple solution is to add the resin to the cloth before placement, i.e., make a "pre-preg." Follow this procedure:

1) Estimate the number of strips required during the working session.

2) Using a marker pen, such as the "Sharpie", mark off a piece of cheap clear painter's drop cloth plastic with as many strips as needed, side-by-side. Turn the plastic over so that the marking is away from the working surface. (The resin dissolves the ink.)

3) Cut one or two pieces of glass cloth a little larger than the work area and place the cloth over the work area.

4) Pour on mixed resin and squeegee it out enough for a rough distribution.

5) Place a second layer of clear plastic over the work area and squeegee to fully wet the glass and drive the excess out at the edges of the work area. With a little practice, the builder quickly learns to fill the glass efficiently with little waste of excess resin.

6) Using fabric shears, trim the plastic around the work area.

7) As needed, accurately cut off each required "prepreg" strip, following the lines marked in Step 1. The strip is now a sandwich of an accurately wet out piece of fabric supported between two pieces of clear plastic.

8) Remove one of the plastic layers and apply the wet strip to the work. Smooth it out, then remove the plastic carrier.

9) Make small adjustments in the positioning of the wet cloth with the fingers or a brush.

This is a very effective technique. It is quick, delivers properly wet out material to tight places, and keeps narrow strips of bias-cut cloth from unraveling.

The method is also very useful with wider strips of cloth. The plastic backing will not permit a double curve, but will provide very good control of fairly large pieces of wet out cloth for an initial placement on a curved surface. After placement, the backing plastic is removed and the wet cloth can be brushed or squeegeed into final position. This will be very useful when making a "splash" mold over a complex surface like a rudder fairing or an engine cowl.

Early sources have recommended the use of wax paper for this purpose. Wax paper has the problems of being fragile and risking wax contamination in a layup.

Hot Wire Foam Cutting

Cutting styrene foam with a hot wire is one of the most elegant techniques in the composite field, and probably the most fun. It produces more useful shaped parts, faster, for the least work than any other method. Used to make wing cores, it is just as easy to make a tapered and twisted swept wing with a variable section as it is to make the simplest of straight "Hershey bar" wings. To cut, a foam block is placed between two accurate templates. A straight wire is heated electrically and drawn through the foam between the templates, melting the foam in the path of the wire.

Hot Wire Saw

The sketch below shows one way to make a hot wire saw. One long saw (6 feet) may be required for cutting sections for a wing, but it is unwieldy for general use, so at least one short one (two feet long) should also be made. Hot wire saws are so simple to make that there is no reason to accept the inconvenience of using an awkward saw. The sliding dowels on the long saw are used as handles for additional wire guidance. Shorter saws do not require the dowels.

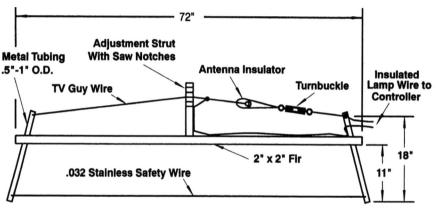

Hot Wire Saw (Typical)

Standard aircraft stainless steel safety wire (.032 or .041 inch) can be used for the cutting wire and works very well except that the optimum cutting temperature is not very far below the temperature at which the wire softens. Inconol non-stretching wire (also .032 or .041 inch) can be used to produce a saw which is a little more tolerant of adjustment. This is probably significant only for the longer saws, but is a worthwhile refinement.

Basic Cutting Technique

Proper adjustment of wire temperature, wire tension, and cutting speed can only be learned by trial and error, however it is remarkable how quickly a builder "gets the hang" of this process when practicing on scraps of styrofoam. The best thing to do is to connect up the wire saw and experiment. Proper conditions are when the wire has a small but noticeable drag in the cut, a faint hissing sound is heard, and fine plastic hairs follow the wire at the end of the cut when the saw is removed. If the temperature is too low, the cut will be slow and the drag will be excessive. If the temperature is too high, wire tension may be lost, there will be no hissing, the cut will be too easy (too big a cut), and the plastic hairs will not result. This is one of those procedures that is actually easier to learn by practice than it is to read about it. More detailed instructions will be given with the example in Chapter 10.

Hot Wire Slicing and Squaring Tool

It is useful to make a simple "baloney slicer", which is a flat board approximately 18 inches by 24 inches, with two vertical steel guide strips at one end. See the sketch on the next page. Odd shaped scraps of styrofoam from the wing cutting operation can be trued up or sliced into thin sections and used in many places where small pieces of core are required. The slicer is handy for producing accurate flat faces for joining blocks of foam before cutting to form. A smaller saw is convenient for use with this tool.

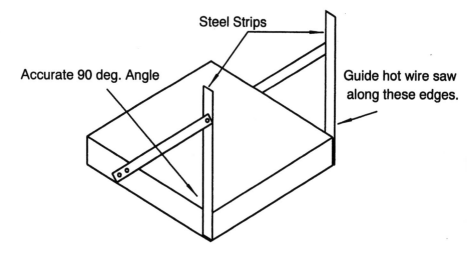

Steel Strips

Accurate 90 deg. Angle

Guide hot wire saw along these edges.

"Baloney Slicer" Foam Trimming Table

Structural Corners

Corners are a necessary evil, but we need to deal with them in many places in an aircraft. From our previous discussion, we see that an infinitely sharp "perfect" corner must produce an infinite stress concentration which no material can stand. We want the load on the corner joint to follow the orientation of our composite fibers to some reasonable degree. We also want to provide sufficient area in the joint to carry the load.

Consider a very long, heavily loaded, railroad freight train. There is some minimum turn radius that the railroad designer must provide if the train is to go around a turn, each car pulling the next. Clearly, a square corner is impossible, so the designer makes the best compromise, choosing a radius so that the thrust of the locomotive, reacting with the side force from the rails, always generates the necessary force to move the following car. The larger the radius, the easier it is to make the turn.

Similarly, where it is required for a skin to go around an outside corner, the designer calls for some minimum radius. At least a 3/16 inch radius is required where medium weight glass is to go directly around a corner. Where the fibers cross the corner at 45 degrees, as little as a 1/8 inch radius can be used. More radius, where allowed, makes a stronger joint and makes it much easier to make the glass stick to both surfaces when wet.

Very stiff fibers, like carbon, are not used over tight corners.

Inside corners are required where a wing rib joins a wing skin or a bulkhead joins a fuselage skin. These are handled as shown below.

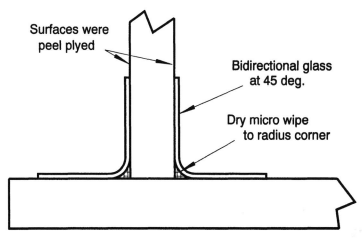

The Inside Corner

The designer will have anticipated this corner joint, so there should be peel ply over the surfaces to be joined. Remove the peel ply and buff with Scotchbrite before assembly.

Provide a corner radius by wiping dry micro in the corner, trying not to get micro where a glass-to-glass bond will be made. Use a curved ice cream stick or the end of the little finger to make the radius. It is a good idea to mask off the joint area with tape to avoid a cleanup problem.

Cut a "prepreg strip" of bi-directional cloth (described above) to the proper length and remove one plastic strip. Fold the prepreg strip lengthwise and work the wet cloth into position. Remove the second plastic strip and dress up the placement with a brush or with the fingers. For a really neat job, some builders finish the joint by putting peel ply over the edge transition, though this serves no structural purpose.

This technique produces a very conservative joint, in most applications. Consider what must happen if this joint is ever to fail: Some concentrated force must occur such that the glass fibers must shear or the very large area secondary bond must fail. Most designers specify a bonding strip about 1 1/2 inches wide, which makes it easy to handle, but produces a bonding area far larger than necessary for the design loads. With this in mind, save weight by making the bonding strips no larger than the designer calls for. "Good measure" here only adds weight.

Where it is necessary to form a sharp corner between two glassed surfaces, a "flox corner" is used to provide the necessary bonding area and stress path. See the sketch below.

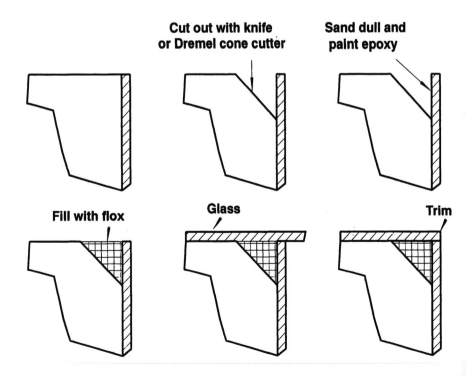

Flox Corner

Repairs

Accidents will happen and errors will be made. Fortunately, safe repairs are relatively easy to make with composite structures.

The basic ideas in a composite repair are: (1) fill damaged core material with dry micro, and (2) sand back damaged laminations to get to undamaged fibers. The general rule is to sand back one inch of ply for every damaged layer. If there are five plies damaged, sand an area five inches wide. This will produce a very large shear area to carry the loads.

Replace the damaged plies with the same layup schedule as the original, overlapping each ply by one inch.

The Knife Trim

About five hours after a layup, depending on the resin system and the temperature, the resin has hardened to about the consistency of cardboard or celery stalks. At this stage, it is firm enough to hold its position, but is very easily cut with a common utility knife. Using a sawing motion, the builder can neatly remove all the messy excess glass fiber from an entire wing in a few minutes, leaving a clean edge that requires almost no additional finishing. If the laminations are disturbed during the trim, they can be pushed back into position, reuniting the "green" B-stage resin. This trimmable property of resin layups is one of the unexpected satisfactions of working with composite technology. An easy knife trim taking seconds avoids hours of dusty, messy, less-accurate work trimming the completely-cured composite.

Finishing - The Basic Problems

We have described here a moldless process which produces an un-finished surface. The hardened composite surface has the inevitable small irregularities associated with hand work and has the surface texture inherent in the fiber surface. The surface must be filled with a very light weight filler so that the contour can be adjusted to the design specifications and so that the surface will have the desired aerodynamic smoothness.

Essentially, the task is to fill all low spots and surface texture with very dry (and light!) micro so that the surface can be sanded down to final specification without sanding into the load-carrying fiber skin. Experienced builders agree that this is much easier to say than it is to do. While it is true that a surface of hardened very dry micro is easy to sand to a precise surface, it is the very devil to handle and spread when wet. The process is to paint a very thin coat of epoxy on the surface to be filled and then squeegee the very dry micro over the area. This is hard to do and takes a while to develop the technique and the patience to put up with the poor spreading char-acteristics of the micro. It can be done, however, as many have proven. The temptation is great for the inexperienced builder to try to improve spreadability by using micro which is too wet (so spreads better), but the price is poor sanding and increased weight.

After the micro fill, the resin matrix is allowed to harden. The builder than inspects the work with steel straightedges or other means, and sands out the surface, usually with long sanding blocks. Problem areas and low spots are identified, new micro is added, the job is allowed to harden, and the task is repeated until the desired surface form is achieved. Many builders believe that this part of the moldless system is the least enjoyable part of the entire project and takes much longer than they ever expect. After several patient cy-cles of filling, squeegeeing out, and sanding, the job is finally fin-ished.

Bob Martin's Method

In 1985, Bob Martin, of EAA Chapter 338, San Jose, CA, demonstrated a technique which offered hope of greatly simplifying the job of filling a cured layup surface. He recognized that what was needed was something that would make the dry micro wet enough to work, but which would evaporate and would not damage the proper hardening of the epoxy. The idea was that the solvent would not react with the epoxy.

He tried isopropyl alcohol, as sold in drug stores as rubbing alcohol. He demonstrated that adding roughly 5% of the rubbing alcohol was sufficient to make the dry micro flow freely. It did not mix well when stirred, but, with persistent stirring, dispersed the micro quite well. He used a *Dragonfly* elevator to demonstrate smoothly filling the entire surface, without high spots, in only a few minutes with a squeegee, scraping off the micro, and doing it again! It was easy to squeegee out and easy to assure that the fill was complete. He had the new problem of having the mixture evaporate *too* fast, requiring that he use a spray bottle of rubbing alcohol to wet it down for handling.

He was understandably concerned that the characteristics of the epoxy binder might be damaged by the solvent, so he made up a large number of test coupons of varying thicknesses, from a few thousandths to about 1/8 inch, and passed them around the room, inviting the members to break them. They had the appearance of very light pumice, with a slight porosity. There was general agreement that, while no one wanted to make a *spar* that way, it certainly made strong enough *paint!* After the session, the solvent had evaporated from the *Dragonfly* elevator, leaving the familiar smoothly filled surface. He displayed another *Dragonfly* elevator which had been previously filled by the method and which had a glass-like finish. Note was taken by the crowd that the originally mixed micro was quite dry indeed and would have been beyond handling by the old method. Builders present who had suffered for weeks finishing *LongEzes* the hard way were deeply moved.

The Fog Coat: A Warning Layer

Bob Martin also made the excellent suggestion that a very thin "fog" coat of colored lacquer can be used to serve as a warning layer for the sanding operation to follow. He found that this thin coat did not affect the adhesion of the micro fill layer.

Most authorities warn sternly that a power sander is NEVER to be used. Bob, however, used a low power orbital sander and found that the use of the colored lacquer layer gave adequate warning to avoid mistakes. He also demonstrated the use of an extruded aluminum channel to serve as a sanding block, which held an accurate line over a length of three feet.

The author suggested that the alcohol might indeed be acting as a true solvent, but noted that rubbing alcohol, as sold, is a mixture containing 30% water. The water could be acting as a dispersant, as oil and vinegar disperse each other when the mixture is shaken. The effect would be the same. Both liquids would quickly evaporate from the very thin layer.

Months later, the author needed to replace an outside door on his house, which was subject to severe weather and sun stress. It was reasoned that if a solvent/dispersant which contained 30% water and evaporated too fast worked so well, why not try 100% water dispersant on the non-flying door? The experiment worked very well and the door in question is good as new after seven years of hard service.

Rex Taylor's Really Dry Micro

In 1985, Rex Taylor was in Eloy, Arizona, where he operated Viking Aircraft, supplier of the *Dragonfly*, and HAPI engines. For a time, he operated a "Fun Fly Center", where builders could come to Eloy for two weeks and use his hanger, facilities, and expertise. The goal was to do all the early work to assemble a *Dragonfly* prefab kit, including the major layups and critical alignments. The author bought a kit and attended a session.

Immediately after laying up the bottom of the *Dragonfly* canard, Rex suggested that we get a start on the inevitable fill job by pouring the pure unmixed micro on the wet surface. He reasoned that the tiny pools of pure liquid epoxy that lay between the woven yarns were essentially wasted and might as well be asked to soak up some microballoons. The experiment was tried on the bottom of the canard where it was covered by the fuselage. Wearing respirators, we poured pure microballoon powder over the wet surface and distributed it with a scrap of peel ply, leaving an excess of dry powder on the surface.

The following morning, we brushed and vacuumed the cured surface to discover that we had made an excellent first fill. More filling would be necessary, but we had made a good start without adding any epoxy at all to this driest of possible micros. That day, we filled the entire upper surface of the canard the same way in a few minutes.

The Smudge Stick

When finishing a surface, high and low spots can be identified with a "smudge stick", a straight board which has been dusted with carpenter's colored chalk. When rubbed on the surface to be finished, low spots don't pick up chalk. Low areas can be marked with a pencil and filled.

Chapter 10
Basic Moldless
Wet Layup Techniques

Building a Foam Core Wing

We will demonstrate the application of most of the basic moldless techniques described in Chapter 9 by building a simple foam core wing section. This simple section will not include an internal spar shear web, so is more typical of an aileron than a full-size wing.

Templates

The wing section is formed by a pair of accurately made templates, as shown below. Sheet aluminum or 1/8 inch painted Masonite makes good templates.

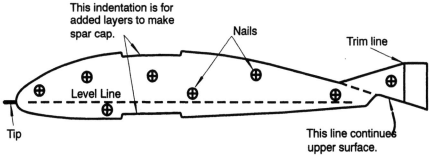

Template Features

The pair of templates is used to guide the hot wire cutting saw in cutting out the foam wing core. Each template is designed to allow for the thickness of the outside skin that is to be added and for the loss of foam due to the cut (about .040 inch). As discussed in Chapter 8, the thickest part of the wing section must carry the greatest bending strain, so additional spar cap material is added to carry that load. Notches are provided as shown to provide room for the added layers.

The tip extension on the template is used to start and coordinate the cutting operation described below. Rather than attempt to include this tip as part of the template, it is easiest to glue a finishing nail to the template after the template is trimmed and sanded to the final dimensions. The presence of the tip leaves a very thin ridge of foam on the cut core which is easily removed later.

Builders should be aware that some plans for foam core structures call for depths of spar cap notches which leave very little room for error; i.e., unless the layup that is to go in the notch is almost perfectly packed, the total layup thickness of the cap may come slightly above the required finish line and require scrapping a half-done wing. It is best to be very sure that the indentation is deep enough. If a thin layer of filler is required to bring the wing to the required surface, that is a small price to pay to save a major part.

The finished wing trailing edge comes to essentially a feather edge and it is not practical to handle such thin unsupported foam. We get around this problem by temporarily providing extra core material at the trailing edge. The thin edge is formed later, after one skin is in place to support the structure. Note the temporary supporting "tail" of extra foam in the drawing of the template.

The level line is marked on each one of the pair of templates, as shown on the drawing, to serve as an alignment reference. Each template is carefully aligned with an accurate level and secured to the ends of the foam block through drilled holes with close-fitting nails. (Be sure that no part of a nail can be hit by the cutting wire.)

The numbers on the templates are reference numbers used to coordinate the cutting operation. See the sketch on the next page.

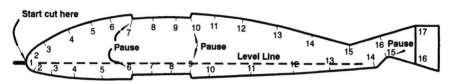

Template With "Calling Numbers"
For Coordinating Hot Wire Cut

The sketch below shows a foam block, with the templates in place, set up for cutting.

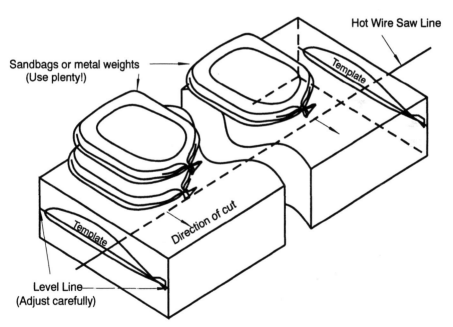

Hot Wire Cutting of Styrofoam Block

Note that the hot wire cutting operation must be completed before core sections are permanently joined with micro. Hardened micro cannot be cut with the hot wire.

When necessary for a large part, separate blocks of foam may be temporarily held together for cutting with duct tape. If the blocks are not exactly the same thickness so that duct tape cannot be used to hold them, small patches of 5-minute epoxy can be used. (Keep the epoxy spots well clear of the line of the cut!) After the cutting operation is complete, the foam sections can be broken apart and the small damaged areas repaired with micro. Final joining with micro across the entire joint is done after the cutting operation.

The foam block to be cut is placed on the bench and firmly and continuously supported by whatever means is convenient. The top of the foam block is held in position by weights. Rutan and the *Dragonfly* like to use old iron scraps for weights, because they are compact, but some builders have found that plastic kitchen garbage bags filled with sand are easier to find, handle, store, and pile up on the work as needed. It is very important to use plenty of weight to assure that there will be no bowing or movement. A typical wing section cutting will require at least 100 pounds of weight and 150 pounds is better.

Cutting Procedure

Before making a cut, the temperature of the saw should be checked by making a practice cut of a piece of scrap foam. The temperature is correct when the proper hissing sound is heard and the cutting wire pulls out of the cut followed by a group of fine plastic hairs. It is also best to rehearse the cut with the cold saw to assure that gross errors are avoided and that the saw will clear the stacked weights and the table top. If additional table clearance is required, the foam block must be fully supported over the entire area. (As always: measure twice, cut once!)

Starting at the leading edge, the leader and a helper each take one end of the saw and move together, cutting toward the reference tip at about 1/2 to 3/4 inch per second. Both cutters announce when the cutting wire is resting on the reference tip about 1/4 inch from the leading edge of the cut. The leader then says "Ready....CUT" and the wire is moved to the leading edge at the reference tip and held for about one second. The two cutters then work together, slowly moving the wire over the top of the template while the leader calls out the number position of the wire. The helper tracks along with the leader, adjusting the cutting position to suit. Wherever there are abrupt changes of direction, such as at a spar cap notch, both cutters pause for a second or two at the corner to allow wire lag to catch up. (This condition is recognized when the hissing sound stops.) After the top cut is complete, the saw is withdrawn and the process is repeated to make the bottom cut.

Basic Moldless Wet Layup

The following description will assume that glass is used for the primary reinforcing fiber. One of the great advantages of glass is that it is transparent when wet and it is very easy for the amateur builder to see and control the wetting process through many layers. The techniques are almost exactly the same for opaque fibers, such as carbon, but there is greater dependence on the builder's skill and experience to assure proper wetting because the work cannot be easily inspected.

Preparation and Minor Repair

The cut foam core is brought to final dimensions by sanding. Where a long straight surface is required, as with a wing, the best tool is a straight sanding block two to six feet long. The foam sands very easily and it is easy to overdo. Inspect frequently with a steel straightedge. Note that a line of hardened micro where two foam blocks are joined does not sand out with the foam. If a line of hard micro exists at the core surface, it is essential to cut the harder material out below the surface before final sanding. This is quickly done with a high speed model maker's hand cutter such as the Dremel. The small groove that results from undercutting the hard micro will be filled with new micro after the final sanding.

Even with ordinary care in the shop, cut foam cores seem to acquire small chips and dents and nail holes. These voids must not be ignored, but are easily corrected by cleaning out and filling with dry micro.

Setup

The lower surface of the wing is laid up first. Before beginning the layup, it is essential to double check that the core is accurately and firmly located on the work bench exactly as the designer has specified. Here is where patience pays. A little hurrying or fatigue may lead to a long repentance. As necessary, the bare core can be bent and twisted slightly with weights and braces to bring it into perfect position.

The best and easiest support is the scrap foam block left over from the hot wire cutting operation, which is an exact fit to the core. For our sample wing, we will need to wrap the skin slightly beyond the center of the wing section, as shown. To allow for access, the supporting core should be cut back about three inches, as shown. (About one minute's work with the hot wire!)

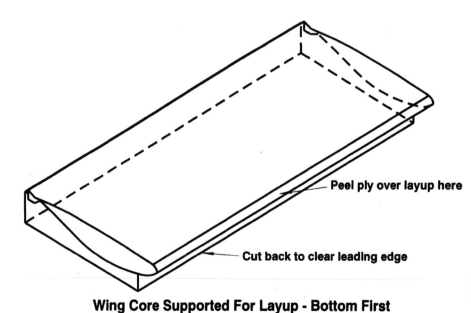

Peel ply over layup here

Cut back to clear leading edge

Wing Core Supported For Layup - Bottom First

For the cases where we will not have the convenience of accurately cut foam scrap to use as a support, the designer will generally describe jig blocks which accurately fit the core and which are used to hold the wing in the specified location. These jig blocks are carefully positioned on the chipboard surface of the bench and held in place with Bondo. If there are corrections to be made, it is very easy to knock off the jig pieces, readjust, and reset with new Bondo. The core can be held in the exact final position with supports made of scrap wood and attached to the foam and to the bench with little patches of Bondo or 5-minute epoxy. (Note that Bondo dissolves polystyrene, but epoxy does not.) If it is necessary to straighten out a bow or twist in the core, this can be done by securing the core to a support made of scrap wood with a patch of 5-minute epoxy. After one surface of the wing has been glassed and has cured and the wing is removed from the bench, there will be a small damaged area where the 5-minute epoxy was broken out. This is easily repaired with dry micro.

Masking the Wing Core for the Layup

The upper wing skin leading edge will overlap the bottom skin in a second layup. We need to provide for a neat edge where the bottom skin ends. To mark the edge, place a strip of 2 inch silver duct tape along the entire leading edge of the foam with the edge of the tape at the leading edge of the core. Press only one edge of the tape on the foam (about 3/4 inch) and let the rest of the tape hang down. Cut wide strips of brown grocery bag paper or newspaper and attach it to the free hanging part of the duct tape to serve to protect the foam from epoxy spills or dust.

At this point, the builder should note that a layup operation, once started, must be completed. Sufficient time, energy, and staying power must be allowed to get the job done in a single session. Time must also be allowed to complete the knife trim. Whether this is an advantage or a disadvantage over aluminum construction can be debated after the session is over and the builder is pouring beer for her helpers.

Filling Foam Core Surfaces

To avoid stress concentrations which could cause failure of a part, it is extremely important that the bond between core structures and skins be strong and continuous. A void-free interface is essential. Styrene foam has large cells which would absorb an excessive amount of heavy liquid resin and/or create the risk of voids (and a poor bond) if the cells were not filled at the skin interface.

The styrene foam surface is filled with "thin micro" which has been mixed to the consistency of thick gravy. The general principle when working with micro is to use the lightest (most viscous) mixture which will do the job. In this case, we wish a mixture which will just flow into the surface cells of the foam when the micro is spread out with a squeegee. Use an excess of micro and spread it fairly thickly. The proper viscosity is shown by the sound of faint crackling as the mixture enters the cells. If the micro is too thick, it will not penetrate or squeegee out smoothly. A little practice with scraps of foam will soon develop the technique. After about four minutes, when the crackling sound has stopped, the excess thin micro can be removed with a squeegee. The finished filled surface should just show the pattern of the cells through the micro.

Urethane foam has very small cells, so a separate filling operation is not necessary. The surface of a urethane core is coated with pure liquid resin, which fills the surface, assures a continuous bond, and also makes it easier to place and wet out the first layer of skin.

Laying Up the Spar Cap Notch

The notch in the wing section has been designed to be just deep enough to hold the required layers of unidirectional spar cap fibers and bring the level even with the rest of the wing core. (The templates shown in the sketch have a rather exaggerated notch.)

It is usually cost and weight efficient to use a high performance fiber like carbon or S-glass in a spar cap because nearly all of the useful load is concentrated in these structures. (Recall Chapter 4!) Many designs call for these fibers. Ideally, the fiber stack will come exactly to the level of the wing core. In practice, it is preferred to allow a little extra clearance space and fill any small error in wing form when the wing surface is finished.

Prepare a batch of thin micro and fill the bottom of the spar cap notch as described above.

Identify the previously-cut uni-directional cloth sections and place the correct cloth section in the prepared spar cap notch. Wet with epoxy and squeegee as necessary to fill the fibers and work out the white areas (air).

It is often possible to obtain the unidirectional spar cap material in the form of tape the same width as the spar cap notch. The use of unidirectional tape is a great convenience in handling. If tape is used, there will usually be a lengthwise red thread at the edge of the tape. Pull out the red thread. With the red thread gone, the cross threads that hold the fibers together are then freed and can be removed, leaving nothing but the straight unidirectional fibers in place.

Add layers of fiber as specified, removing the binding threads in each layer (if any), and rewet and squeegee until the fibers are neatly wet out and any excess resin has been squeegeed out and removed. At this point, the spar cap notch should be filled. Fill any gaps between the edge of the notch and the side of the stack of fibers with dry micro. (But don't contaminate the resin in the interior of the spar cap with micro!) The wing skin which will be added later will be bonded to the spar cap and will add somewhat to the strength of the spar.

Laying Up the Wing Bottom

Before beginning the layup of the bottom wing skin, recall that we will need a bond joint with the future top skin at the trailing edge. We prepare for this by placing a strip of peel ply on the trailing edge of the core as shown in the sketch. See that this surface continues the line of the upper surface of the wing and will be at the bond line. The peel ply will be the first layer under the bottom wing skin and we will lay the glass skin over it.

Place the cloth layers on the wing core as specified by the designer. For large sections, such as when skinning a wing, two people are required for this operation. Usually, one layer is applied and wet out at a time, as specified by the designer. Several pieces of fabric will usually be required to cover the desired area. Overlapping, which will cause a bump, should be avoided. For cloth layers, adjustments in position can be made by pulling at the edges of the cloth and it is sufficient to bring the pieces very close to each other without overlapping. For unidirectional layers, the pieces can be worked quite close to each other by allowing a slight wave in the line of the fibers. (Small variations from perfect orientation are OK.)

If using multiple layers, such as stitch-bonded triax, instead of woven cloth, wetting out is helped by applying extra resin to the surface before placing the fabric. Care must then be taken that the excess resin under the fabric be well squeegeed to bring it up into the fabric and not left on the surface of the core. Stitch-bonded fabrics cannot stand enthusiastic squeegeeing as well as woven cloth without disturbing the fiber orientation. This problem is avoided by placing a section of the cheap clear polyethylene plastic sold for painter's drop cloths over the work and squeegeeing through the plastic. Use a piece about two feet square and peel it back as needed when pouring on more epoxy or moving to another location.

> In the author's Dragonfly, all of the specified layers were applied at once in a single giant piece of triax, a great saving in work. Wetting out the three layers was not a problem in an $85^{0}F$ hangar in Eloy, Arizona, in August!

Some resin will be seen soaking the dry white cloth immediately after placing the cloth. Add resin by pouring it over the cloth and squeegeeing it out. Here again, a little practice will make the process clear. The idea is to wet the fabric just enough to completely fill it without allowing white areas (air). Any extra resin beyond full wetting only adds weight and costs money. It is convenient to make the first layer a little wet to assure a perfect bond to the core and to make wetting the next layer a little faster. The squeegee is the main tool used for spreading resin and removing excess. "Stippling", or stabbing downward with a paintbrush, is sometimes useful to bring up excess resin or when applying tapes to inside corners. (Some builders modify the stippling brush by cutting off about half the bristles of a cheap throwaway paint brush with scissors.)

Additional layers of fabric are added to complete the layup as specified by the designer. It is convenient to work with the lower layers a little wet and bring the excess epoxy up into a newly applied layer.

Finish the layup by applying peel ply over the leading edge bonding area shown. If peel ply tape is being used, it may be necessary to use two strips to cover the required area. This will be removed after the layup cures.

After the layup is complete, each layer should be scissor trimmed to approximately 1/2 inch of the edge to prevent overhang which might cause a void. If too much excess cloth is allowed to remain, it tends to wet with excess resin, get heavy, hang down, and make a void at the bend.

The Knife Trim

Use the knife trim to remove excess fabric along the entire trailing edge and the tips. Knife trim along the leading edge, following the line of the silver duct tape. To do this, it will be necessary to cut very slightly into the foam core, but this cut will be filled when the top skin is made. Smooth back any irregularities with the gloved hand.

Filling the Trailing Edge

See that we have deliberately formed a gently curved indentation at the trailing edge below the line of the wing surface. The wing skin follows this curve.

Fill the indentation with dry micro so that the level of the fill comes ABOVE the line of the wing surface. This excess will be sanded level after the layup has cured.

Curing Between Layups

After the knife trim and filling the trailing edge, allow the layup to cure hard. Do not rush this phase, because we set the core up very carefully before the layup and are depending on the new skin to hold the shape until we finish the wing. How long is safe to wait will be determined by the resin system used and the temperature.

Laying Up the Top Skin

While the wing is still supported for the bottom skin layup, sand off the excess micro in the training edge indentation so that it is flush with the wing line.

Remove the silver duct tape and remove the wing form from the support used for the bottom layup. Turn the wing over and position as necessary for the second layup. If the scrap foam produced during the hot wire cut is available, it may be possible to use that to position the wing. In any event, the wing must be accurately set up on the work table by whatever means the designer recommends. Even though the bottom skin has hardened, very small adjustments in wing form are still possible and should be made, if necessary.

The wing should be firmly attached to the work bench with fixtures and braces using Bondo as necessary. The builder should be very cautious in this set up and should support the work generously, assuming that he or a helper will accidentally kick the table during the pressure of completing the second layup.

Remove the peel ply from the leading edge to expose the clean bonding surface for the new skin. Sand off the blunt edge of the lower skin so that it tapers smoothly to the surface of the foam.

See that the new lower skin and micro fill now makes the trailing edge of the wing stiff so that the extra foam reinforcing tail is no longer necessary. Use a hack saw blade to remove most of the foam tail just above the line of the wing surface, as shown below. Cut just above the peel ply layer applied when the bottom layup was made.

Pull off the peel ply which was laid up under the skin during the previous layup. This will break off the thin piece of attached foam and expose the new bond area for the trailing edge. Sand the foam core to blend smoothly with the exposed trailing edge bond.

Inspect the wing carefully and repair the inevitable dings and chips with dry micro. Repair any damage due to the restraints needed by the first layup. Be sure that the setup is perfectly positioned and, if it isn't, make it so.

Lay up the top spar cap and the top wing skin as described above. Note that the knife trim technique cannot be used where the top skin overlaps the bottom skin at the leading edge because we cannot cut into the bottom skin. Trim the wet upper skin with scissors so that there is an overlap of about 1 1/2 inches and smooth this wet skin into position. Cover the somewhat ragged edge of the wet skin with a layer of peel ply. This will produce a smooth transition without the need to cut into the fiber.

Complete the knife trim and allow time for a good cure before removing the wing from the supports.

Finishing

The basic principles of filling the textured fabric surface were discussed at the end of Chapter 9. Low areas should be identified by inspection with a steel straightedge and a smudge stick and then filled with micro and sanded out. Repeat as necessary. As the process converges toward the final surface to be painted, finer sandpaper will be used.

Chapter 11
Hollow Structures

Alternatives to Solid Cores

While followers of Rutan might use mostly-solid wings, we still want to place fuel tanks, landing gear, and other hardware inside the wing surfaces. We would certainly find a solid fuselage to be inconvenient. We showed previously how useful structures resulted if we used load-carrying skins over a core which could carry shear forces between the skins. If we remove that distributed core to insert a fuel tank, for instance, we lose that necessary shear-transferring means and the structure is no longer a useful beam. In this chapter, we will consider ways to extend composite construction techniques to make hollow structures.

Unlike the designer of an aluminum aircraft, the designer of a composite aircraft has a unique choice between four options: A section of the aircraft may be:

(1) full-thickness core sandwiches, as in the Rutan-type airfoils discussed in Chapter 8,

(2) self-supporting single skin, where there is a significant short-radius double curvature, as in cowls, wheel pants, and wing tip fairings,

(3) extended gently-curved surfaces where the surface is a thin core sandwich form, as in fuselage sections or wing skins, and

(4) surfaces made from locally stiffened single skins.

The composite designer may combine these options of skin sections and core sections into large assemblies with complex shapes.

Self-supporting Curved Surfaces

A piece of paper bends freely from its own weight, if picked up from a table. Similarly, a flat sheet of aluminum, steel, or plywood has little useful strength if used as a bent beam. If these flat materials are bent to form a single curve, such as a tube, they produce many useful structures. The material becomes stiff along the axis of the curve. A good example is formation of a boat hull from four big sheets of curved plywood bent over a simple frame. Two sheets make the V-bottom and two sheets make the sides. A round sheet metal stove pipe is another example. An aluminum sheet which is curved slightly around an aircraft fuselage and riveted to a circular frame is another.

A double curve, such as a bowl, is even more efficient. An eggshell is an efficient double-curved structure which develops surprising strength from its fragile material. A stamped steel automobile fender needs very little additional support beyond its own form. Formed into a double curve, sheet materials become very useful as engine cowls, auto body parts, or sections of the WW II wooden Mosquito bomber. The typical internal auto body steel stamping is full of domes and ridges which stiffen what began as a large sheet of flexible flat steel. A typical small fiberglass boat, which often uses large unsupported skin areas, will be seen to have few areas which are actually flat; there is nearly always some double curvature so that the designer can avoid the use of internal frames. Many small production airplanes use highly curved fiberglass engine cowls or wheel pant covers which are unsupported and made in a mold the same way as parts of a small boat. The common factor in all these forms is that they have a single skin layer, a substantial double curvature, and little need for additional internal structure.

A fabric-covered airplane is the conceptual opposite; the entire surface is completely determined by the internal supporting structure and each area of unsupported fabric shrinks to occupy the minimum surface area. Fabric cannot carry compression. Fabric-covered airplanes, as a consequence, have a great many parts, fastenings, and stitches and require compromises to work around the properties of stretched fabric. Where there is a great deal of curvature, or air loads are high, additional support and/or attachment must be provided. (Some WW I airplanes had to shed fabric on the upper side of their wings in a dive before this was adequately appreciated.) Where air loads are small and the surface to be covered is flat, as on the side of a WW I fuselage, a large unsupported area of fabric is satisfactory. As late as WW II, we see that designers of high performance metal fighters chose to use control surfaces that were fabric-covered.

The Locally-stiffened Sheet

We have discussed the flat two-skin core sandwich. An alternative is a flat single-skin sheet which has been stabilized against buckling and stiffened in bending by the addition of one or more stiffening ridges. See the sketch below, which shows a relatively long thin flat surface with a single thin skin which has been stiffened with an attached beam, or "stringer". This is similar in function to the use of "hat section" stringers to stiffen thin aluminum sheets in conventional aircraft.

A flat sheet of composite material - say glass - is laid up on a smooth mold surface, a section of core material is bonded to the sheet to form the ridge, and more skin material is laid over the en-

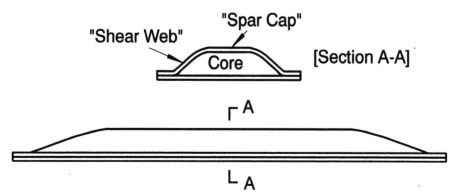

Stiffening Stringer for Single Skin

tire assembly to complete the structure. See that the stringer is essentially a beam, where the top of the stringer is the "spar cap" and the sloping sides serve as "shear webs". The sloping sides are necessary to allow the smooth layup of cloth reinforcing materials. (Many modern skis are built this way.)

We will have more to say about actual construction techniques in later chapters. For now, an example of a practical construction for such a stiffened plate might be three flat layers of glass cloth, laid on a flat form, with a tapered foam core for the stiffener bonded to the surface. Three more layers of cloth are then laid over the assembly. Satin-weave cloth will lie smoothly if the edges of the foam core are rounded gently.

In the sketch below, we show a much larger flat surface with several attached stringers, including one which has a relatively large area. This example might be a "bottom" view of a practical plate where the builder requires a smooth surface on one side. This technique can use less material than a uniform flat sandwich form. The ridge can be as deep as is convenient (therefore locally stiffer) and can be located only where necessary. See that the "large area stiffener" essentially converts the covered area to a conventional sandwich structure in that area.

The stiffened sheet above must be supported at the edges in a practical structure. Conceivably, this sheet/plate might be used as part

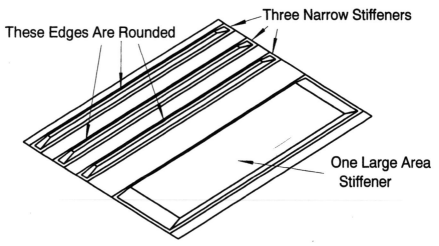

Three Narrow Stiffeners

These Edges Are Rounded

One Large Area Stiffener

Single-skin Plate With Stiffeners

of a wing skin. The point is that we have avoided the need for a very large area formed sandwich.

This method of making a strong plate has the considerable advantage for the home builder that expensive tools or techniques are not required. Many of the advantages of production sandwich structures can be retained. Many variations are possible. For example, performance of some stiffeners might be increased by including carbon fiber tow material to serve as a "spar cap" at the top of a stringer core. It is also possible to adjust the skin thickness as needed, reducing skin thickness in areas of small loads.

Burt Rutan, in his classic book *Moldless Composite Sandwich Homebuilt Aircraft Construction*, gives a wonderfully clear description of the construction of a stiffened plate "confidence layup" to illustrate the strength achieved with the method. He has the builder make a small flat sandwich plate 16 inches by 2 1/2 inches with a 1/2 inch high stiffening ridge about 1 1/2 inches wide. The core material inside of the stiffening ridge is foam. After curing, the builder is asked to try to break the piece over a broom handle by pressing very hard on the ends. It doesn't break. It's a strong part! Nothing makes a believer quite like this demonstration.

> *Years ago, an early VariEze builder, a colleague of the author, brought his Rutan "confidence layup" to work. A number of amazed engineers spent the better part of a noon hour jumping up and down on this light little home-built structure. The strange idea of building a whole airplane this way suddenly became very reasonable.*

The Hollow Wing

In a practical hollow wing, the shear-transferring function must be provided by some kind of a spar system. The spar caps and shear web may be curved around any voids, such as fuel tanks, as necessary. The skin must be supported to serve its aerodynamic function and air loads on the skin must be collected and carried to the spars. This requires that there be some arrangement of ribs or some other support to shape the skin and to carry loads to the spars. We saw how the sandwich structure can provide such an inherently supported skin.

A conventional aluminum wing uses a frame of spars and ribs to support and form the aluminum skin. Like in a fabric-covered aircraft, the designer uses as many ribs or stringers as necessary to give the wing its shape and to collect the loads. The designer must consider the loads to be carried by the skin. Wherever there is a compression load to be carried, the designer must provide adequate stiffness against buckling the thin skin and must add whatever stiffeners are needed to make it so. (Older pilots will recall the wrinkled look of the upper wing surface of the old piston transports as the thin skin deformed under the compression load. This buckled skin wasn't helping the spar much!)

The composite designer might look at the general plan of a conventional aluminum wing design for a small aircraft and consider just duplicating it, using an appropriately formed single layer composite sheet wherever sheet aluminum was used in the original. The composite designer would have to bear in mind that glass composite, for example, is less stiff than aluminum, but the designer would have the convenience of being able to adjust the thickness of her composite sheet as needed. The resulting design would differ in details from aluminum but be entirely satisfactory - probably a little lighter and a little more flexible than the same design in aluminum. It would, however, require approximately the same number of pieces and attachments as the aluminum wing and would be a terrible lot of work. (It would, of course, be spared all the rivets!)

Molded-skin Structures

Sandwich Skin Wings

Current composite aircraft designs make extensive use of formed sandwich structures to span large areas. The Lancair 320 wing, for example, uses only six ribs, one main spar, and a secondary spar. The skins are precision formed curved sandwich structures which can carry large distributed loads to the few strong ribs. The actual load-bearing skins are relatively thin, compared to the total wing thickness, *but the sandwich structure and the curvature stabilizes them against buckling in compression so that the structure is greatly simplified, while being very strong.*

A kit manufacturer may use curved honeycomb or curved foam material to make a large area part.

The assembly of a wing composed of precision-formed sandwich skins is much like the assembly of a snap-together toy plastic airplane. Typically, the top surface sandwich is placed on a table and accurately placed in a final position, then held temporarily by Bondo or other means. The few spar and rib assemblies are then bonded in place. The resulting partial wing is a stiff, accurate structure open at the bottom for installation of whatever else is required in the wing. The bottom skin is then bonded on and the wing structure is complete.

The sketch on the next page shows a somewhat representative cross section of a possible sandwich-skin wing. The sketch suggests only two of the many combinations of sandwich skin and spar assemblies that are possible. In this case, the forward spar is shown as two precision-molded "C" sections which are assembled facing each other, with spar cap material laid between the spar and the skin. The spar molding, then, constitutes the shear web and the builder would provide the separate spar cap layup as specified by the designer. (Alternatively, the spar cap could be a separate molding or could be integrated with the web.)

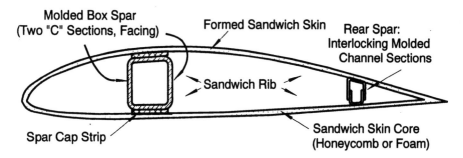

Molded Box Spar
(Two "C" Sections, Facing)

Formed Sandwich Skin

Rear Spar:
Interlocking Molded
Channel Sections

Sandwich Rib

Spar Cap Strip

Sandwich Skin Core
(Honeycomb or Foam)

Molded Sandwich Skin Wing Section
(Skin Thickness NOT Exaggerated)

The rear spar example is shown as two interlocking molded chan-
nels which do *not* use sandwich construction. (Recall that one purpose of sandwich construction is to avoid buckling of one skin
under compression loads. If the column length of the skin is short
enough and sufficient cross-sectional area is provided to carry the
compression load, then it is not necessary to go to the sandwich.)

The sketch does not show a specific method of joining the upper
and lower skins. Usually, the skins are tapered over the core to
form a joggle joint with sufficient surface for the bonding adhesive.

From the viewpoint of a kit builder, this system is a huge saving in
work. There are very few parts and the parts are largely self-aligning. There is very little surface preparation for finishing. The parts
carry the accuracy and surface finish of the precision molding tool
used by the manufacturer, so the builder does not need to build a
mold or an assembly jig. Building a precision mold and obtaining a
very good surface finish for that mold is considerably more work
than building the actual structure.

Locally-stiffened Skin Techniques

The high-temperature technology used by a kit manufacturer is essentially not available to the individual kit builder. The home builder of a single airplane will probably not wish to invest in curing ovens or storage facilities for prepreg materials and will not wish to assemble reliable curved honeycomb core sandwiches. The complexities of curved sandwich core bonding are acceptable in a production shop but are rarely attractive to an individual, even though a superior product is produced.

It is possible for the builder to closely duplicate the functions of the molded parts made by a production kit builder by substituting locally stiffened single-surface skins for sandwich skins. The resulting structure is analogous to conventional single-skin aluminum construction, where the aluminum is stiffened over long flat spans by stringers and ribs. The penalty is that the builder must now make some kind of a mold to shape the skins. (Mold-making techniques and alternatives will be discussed in Chapter 12.) For now, we can assume that a carefully formed smooth aluminum sheet can be used as a female mold.) The sketch below shows a possible single-surface wing design which uses stringers and area stiffeners instead of continuous sandwich skins.

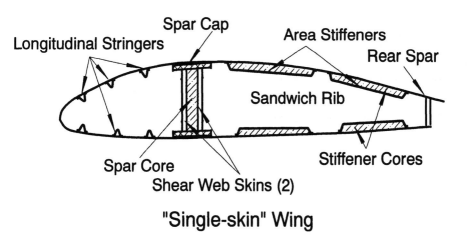

"Single-skin" Wing

This system allows the builder to make an accurate large area surface skin with multiple stringers at a single layup. The technique of making locally-stiffened composite skins is discussed in Chapter 13.

In the sketched example, the main spar is shown as consisting of a spar cap laid up directly on the outer skin with two shear webs laid up on a foam core as a separate assembly. The shear web is then bonded to the cap to complete the spar. See that most of the shear stress is carried in the immediate vicinity of the shear web skins, so the designer must be sure to provide enough bond area to carry that stress. An alternative construction would be to curve the shear web skins over the core, as in the Rutan-type wing described in Chapter 10, which would also produce a generous bonding area.

Assembly of the complete structure would be approximately the same as with the corresponding production sandwich part: The somewhat flexible top skin would be accurately held in position on an assembly table while the few spars, stiffeners and ribs are bonded in. After the bonding operation has hardened, the wing shape will be determined and ready for installation of whatever hardware is required inside the wing. Alignment of the wing is then carefully checked and the bottom skin is then bonded on. (*The designer should be sure that adequate inspection holes are provided in the bottom skin to allow inspection of all of the bond lines and/or access to installed hardware.*)

Compression Forces in the Skins

Recall that the ultimate reason we use a sandwich construction to make *any* wing skin is that we must deal with compressive forces. If there is to be no compressive force in an area, we might as well cover it with fabric! The large area sandwich skin, as made by a well equipped kit manufacturer, solves this problem by making the skin uniform. Looking back at our simple truss discussed in Chapter 2, see that it is only necessary to prevent buckling between the *nodes* of the truss. (That is why we showed all of the compression members as if they were stiff tubes or shafts with rings on the ends.)

In a wing made by using locally stiffened skins over a series of ribs, see that the intersections of the ribs with the skins would correspond to the nodes of the truss of Chapter 2. The designer of the locally stiffened wing could vary the stiffness any way she wanted by adjusting the stiffening structures to suit. The skin need not be a sandwich where it is supported by the ribs. This method of making a wing skin offers the opportunity to save weight by using less material in areas of low stress and optimizing for strength where it is needed.

A Comparative Design Study

Hal Loken and Martin Hollman, in their book *Designing With Core* (1988) report a very interesting design study by a major aircraft manufacturer for a rudder for a large transport. Five cases were considered:

(1) a Kevlar honeycomb hollow sandwich skin with relatively few ribs (area stiffened by sandwich, as on page 9-9),

(2) a Kevlar single-surface skin (no sandwich) with many ribs,

(3) a Kevlar skin over a full depth honeycomb core and essentially no ribs,

(4) a graphite skin (no sandwich) with many ribs, and

(5) "conventional" single aluminum skin with many ribs.

The calculated weights came out in the order listed, with #5 ("conventional") being over twice the weight of #1!

Extreme Limits - Human-powered Airplanes

The desire to solve the problems of human-powered flight has given rise to some of the most elegantly engineered "homebuilt" aircraft ever built. Dr. Paul MacCready and his associates became world famous outside aviation circles for their success in winning the two Kremer prizes for human-powered flight. The first Kremer prize was won in 1977 by the *Gossamer Condor* for completing a one mile course. The second prize was taken by the *Gossamer Albatross* only two years later for crossing the English Channel. Both aircraft were well-named as "gossamer" because they were basically extremely light framed aircraft covered with clear Mylar film. The first *Condor* used mostly aluminum structure; the *Albatross* went to a mostly carbon composite design.

Having done so well with human-driven ultra-low-powered aircraft, Dr. MacCready and his group went on to build the *Solar Challenger*, which was powered only by the electric power (5.7 hp) generated by self-carried solar cells. This aircraft successfully flew from Paris to London. The *Challenger* was clearly a design derivative of the *Condor* and the *Albatross*, because it used many of the same techniques of producing and connecting ultra light composite tubing and frame members and it was also covered with 4-mil Mylar.

In 1988, the human-powered record was taken by *Project Daedalus*, using a world-class bicyclist/pilot and a roughly similar ultra-light aircraft to cross the 71.5 miles from Crete to Greece in nearly four hours of continuous human effort.

A World War I aircraft engineer would be mightily impressed by the high performance materials used by the designers of these human-powered aircraft, but he would recognize the familiar philosophy of a light, externally braced frame covered with a very light material (Mylar) loaded in tension. The sandwich techniques we have previously discussed are used only in the construction of spars and the larger frame parts, not in the external surfaces.

The *Raven* project is a volunteer development project in the Seattle area with the intention of human-powered flight 100 miles, in about five hours, from Boundary Bay, Canada, to Seattle over Puget Sound in 1996. *Raven* will use an extraordinarily light true sandwich material of carbon skins over a foam core. This sandwich material will weigh less than 0.5 oz/ft^2, but, unlike Mylar sheet, is a rigid material which can take compression. *Raven* expects to build a wing with a span of 115 feet and an area of 350 ft^2 that weighs only 45 pounds. The fact that this project can even be considered is a remarkable testimony to the potential of sandwich core techniques.

Molt Taylor's "Paper" Airplanes

Molt Taylor was one of the pioneers of the homebuilt aircraft movement. He is justly famous for innovative design and imaginative use of unusual materials. He has designed very interesting small pusher aircraft and has made important contributions to making very long composite propellor shafts practical for small aircraft. Of interest to us here are Taylor's designs which use paper as a core material.

We will recognize that "paper" is just another word for "cellulose felt." Ordinary paper, by itself, isn't very strong, but we saw in Chapter 1 that individual cellulose fibers (cotton) measure slightly stronger than the legendary spider web. Common paper, then, may be seen as an accurately formed thin sheet of cellulose fiber material that lacks only a proper matrix to make an excellent composite.

Molt Taylor's *Micro-Imp* and *Bullet* designs use the type of Kraft paper ordinarily used for grocery bags, but thicker. The single-thickness paper is sold in the industry as *90 pound pulpboard* and the thicker laminated form of the same material is sold as *laminated container board*. With Taylor's method, the builder works with the plain paper sheet stock to form his part - cutting, bending, and cementing the parts as necessary. When the assembled bond has hardened, the builder soaks the paper with a matrix resin to form the final composite. This technique is extremely easy to use and permits fairly large parts to be made without a mold or with the simplest of locating means. (A Boeing engineer might see this as a *very* elementary form of resin-transfer molding!)

In some applications, the single layer of resin-filled paper composite is sufficient. Where additional strength is required, the resin-filled paper is used as a sandwich core by covering the assembled part with 6 ounce fiberglass. Large unsupported areas are stiffened by stringers, as described previously. With Taylor's method, the stringer is a simple triangular ridge of paper which is glassed over with the rest of the assembly (like the stiffened skin example on page 9-3) which produces a hollow triangular core-less tube similar to the "hat section" used in aluminum construction. A hat section is an inefficient form, whether in aluminum or composite because only the top of the "hat" ever sees a critical stress. The advantage here is great ease of construction. Where right angle joints are required, as joining a wing rib to a skin, a fillet of wood or filled matrix resin is used and the glass skin is applied over the fillet.

The builder may make his own "flat sandwich stock" by laying up glass and paper into a sandwich, useful for making ribs. It is also possible to wet the paper core with water, then form the paper to a desired compound curved shape. After the core dries, it can be soaked with resin and the outer layers of glass added to complete the assembly. The wet paper "core" can be formed easily, compared to - say - a thin slice of foam core material. The resulting sandwich part would be lighter than the same part made as single-layer fiberglass, though, of course, more work.

Compared to other possible sandwich core materials for the same thickness, resin-soaked cellulose is rather heavy and over-strength when used with glass skins, but it certainly is convenient and cheap as used by Taylor. It is particularly suitable for small, light structures like ailerons, doors, and fairings. The fact that the paper core can be softened with water, formed to a complex compound shape, then dried after forming, allows true sandwich construction where other core materials might be less practical.

The technique has not "caught on" in the composite homebuilt community to the extent that it probably deserves. It should be considered where the builder might otherwise use a formed soft aluminum sheet or might use several layers of heavier glass without a core. For some applications, like inspection panels, very light 1.45 ounce fiberglass would make a quite satisfactory skin over the "paper" core. It can be used in some applications as a direct substitution for thin sheet aluminum, with the advantages that it will not corrode and can be bonded securely to other composite structures. Essentially, Taylor has shown that "thick cellulose felt" makes a very practical core material where a thin, formable core is desired.

Hybrid Designs - The Role of Tooling

One possible aircraft design philosophy is to identify the few heavy load points of a small aircraft (engine mount, landing gear, fuel tanks, wing attachments, and useful load), recognize that these locations are all close together, and design the simplest possible frame to carry these loads. This small internal frame, or "cage", then has only a structural function, without any aerodynamic purpose, and can be an efficient design. The "external parts" of the airplane, including a lightly-loaded skin, can be built by any means and attached to the strong "cage".

Two very well known and respected designers in the homebuilt community, who have designed several popular all-composite aircraft, have completed very successful high wing designs which chose to use the steel tube "cage". Martin Hollman's *Stallion* is a big *Cessna 206*-like utility hauler and Stoddard-Hamilton's *GlaStar* is like a completely redesigned *Cessna 152*. The *GlaStar* was an instant success, with many kits sold as soon as they became available. The *GlaStar* is particularly interesting because the wing is of conventional aluminum construction (including the use of hat sections!) while the fuselage skin and tail is composite. The reasons for the appeal of the *GlaStar* are clear: It is a good-performing aircraft from a manufacturer of solid reputation, which doesn't take too long to build, all at a good price. Most builders probably don't care too much about being overly pure about all-composite construction; they look at the "bottom line" - the airplane for the money and the work.

A very important difference between a manufacturer and a home builder is that the manufacturer expects to build many copies of a design and the home builder expects to build only one, or at most a very few. The manufacturer then gives much attention to the design of tools and jigs which the home builder could not even consider. Part of the success of the Rutan designs is due to the fact that his technique was *moldless* with an absolute minimum of setup. The builder built only the part, not first the tool and then the part, so the *total* work required was acceptable to the builder.

But the manufacturer *has* tooling. In the case of the *GlaStar*, the kit buyer receives the aluminum wing parts precut with some holes predrilled on a wing jig; the predrilled holes serve to help align the parts and the builder gets the direct benefit of the manufacturer's jig. Without the tooling, the total work and the risk of error would be much greater. One might argue that an all-composite wing might be "better", but the counter argument is that the tooled metal wing is good enough, quick enough, and the price is right.

The "cage" is also quickly built on the manufacturer's precision jig. Engineering design is greatly simplified, particularly for a high wing aircraft. The cabin skin then can be considered to be a light fairing with no load-carrying requirement and ignored as a part of the structure. Sample steel tube cages can be quickly designed, built, tested to destruction, and modified as needed without destroying an entire test fuselage section. In production, the builder gets an accurate pre-built cage and doesn't have to do the expert welding and alignment. The skin fairing is a light factory molding which attaches to the cage.

An all-composite design which didn't use the cage would require that the entire cabin structure be built and tested as a whole, a slower and more expensive undertaking in engineering. The all-composite design requires that all necessary load-transmitting structures, reinforcements, attachments, and bushings be integrated into the design. From the viewpoint of the home builder, this appears as the added layers of fiber which the designer calls out in her design or appears as the load bearing sections in a pre-molded kit like a *Lancair*, a *KIS*, or the earlier "non-cage" *Glasairs*.

The Flutter Problem

Flags flutter. The rapid oscillatory motion, driven by the airstream, is expected of flags but is very bad for aircraft. Flutter occurs when any part of an aircraft moves so that the resulting air loads tend to increase that movement, further increasing the air loads, generally leading to extremely rapid destruction of the part.

Flutter is always associated with some natural resonant frequency in the aircraft structure which is excited by some driving force associated with the airstream. An analogy is the demonstration where a singer can break a nearby wineglass by exactly hitting the natural frequency of the wineglass, or a parent can time the push of a child's swing to make the swing rise higher.

Flutter in an aircraft, as with any resonant structure, can be prevented by changing the natural frequency of the structure or by damping the oscillation. With the wineglass demonstration, water or sand in the glass or a wad of chewing gum on the edge will detune the glass and shift the resonance. Similarly, if the child on the swing drags his feet at each pass, the swing will not rise.

Flutter can occur with almost any part of an aircraft, but the greatest risk with small General Aviation aircraft is associated with the control surfaces. We see that it is almost the general rule that the control surfaces of most small aircraft use counterweights and/or control area forward of the hinge line to balance them at their hinge lines to prevent flutter. The prediction of flutter in a structure, like a tail or wing, is very complex and is definitely not a task for the amateur. It involves the interaction between the airstream and the natural frequencies of all parts of the structure, not just the control surfaces themselves. Commercial aircraft companies employ aeroelastic specialists who monitor a design in progress to prevent flutter.

Throughout this book, the author has warned the builder of the dangers of deviating from the designer's TESTED design. The danger is particularly great when dealing with flutter. An "improvement" may, indeed, make a part stronger, but it may also change an important structural frequency and lead to devastating flutter when the "improved" plane is flown. Get professional flutter analysis.

Flutter Prevention Methods - How Flexible?

One flutter mode of importance to small homebuilts is where the wing twists, driven by the aileron. Consider that an aileron moves down, generating an upward force on the hinge. The hinge force then causes the wing to twist opposite so that the angle of attack of the wing tip is reduced, causing the wing tip to drop, carrying the aileron, which increases the angle of attack of the aileron. If the entire system of twisting wing and aileron is resonant, like the wine glass example, the system will oscillate until something breaks. A dangerous variation of this mode is the case where both wings twist as a pair, with one wing twisting clockwise while the other wing twists the other way. Flutter in this mode might be prevented (detuned) by balancing the ailerons in the two wings slightly differently and/or by building the wings with a slight difference in stiffness distribution so that their resonant frequencies are different.

A "brute force" way to prevent this flutter is to make the wing very stiff in torsion, i.e., have such a high natural frequency for twisting, that no normal flight condition can excite a twisting oscillation. The Rutan wing designs, and those following, like the Dragonfly, make use of multiple layers of glass cloth with the fibers laid at 45 degrees to the span to resist twisting (high stiffness in torsion). This is a necessary compromise in the use of material, because the wing skin then carries normal bending loads less efficiently than if the fibers ran exactly span-wise, hence must be a little heavier. Designs that use sandwich skin wings, like the Lancairs and the KIS, are extremely stiff in torsion.

The ideal wing form for stiffness in torsion would be a hollow tube. There are formulas in the engineering literature for the optimum angle at which to lay fibers when winding a high performance tubular drive shaft, as for a helicopter. (It turns out to be a shallow angle.) Whatever its virtues for a drive shaft, the tube is not our choice for a wing section. Wing designers get torsional stiffness by placing fibers at 45 degrees to the span. With this compromise, they get both bending and torsional stiffness. The spanwise unidirectional fibers contribute little to torsional stiffness.

Flexibility In Composite Designs

Other things being equal, a composite design may be less stiff than a comparable design in aluminum, so the design conditions for for flutter safety margin will be different from an aluminum design. This is not necessarily a bad thing as long as the flutter problem is understood and a proper design made to avoid it. Moderate flexure under load is not a problem. We shouldn't really care if the wingtips of our small homebuilts rise up a few inches in a hard turn; after all, world-class sailplanes have the wingtips rise as much as eight feet! Internal wing hardware like aileron controls and fuel tank mounting should make adequate allowance for wing bending even where the bending is small. As small aircraft pilots, we have grown used to the apparent granite rigidity of small aluminum airplanes, but we don't have to insist upon it.

The B-47, the first all-jet strategic bomber, had extremely flexible wings which drooped down visibly when the aircraft was parked. A visitor could reach up to the wing tip, shake it, and see and feel the surprising movement. The weight penalty of making the wing as stiff as was typical of the earlier piston bombers would have been prohibitive to the mission of the aircraft, so the success of the design depended on successfully designing out flutter. Years ago, the author had occasion to meet the Boeing engineer in charge of this effort and was pleased to see that he had a commemorative award on the wall of his office for this important success.

Whether designed for safe flutter margins, or just from general conservatism, most current composite kits are grossly overdesigned for their stated functions and published flight envelopes. The author finds that the manufacturers will never admit to an outsider what their actual tested flutter margins are. Typically, it would be very hard to break one of the current popular designs by any flight loads below the maximum rated speed. It is interesting to speculate on what might result from a serious weight reduction program done for current molded-skin kit aircraft. There would probably be a moderate reduction in weight and some reduction in material cost, probably cancelled by the fact that the additional engineering and testing isn't free! In a society full of lawyers, the kit manufacturers like that generous margin in crash-worthiness and tolerance of possible second-rate assembly by a builder. If the time should come when a pilot must put down in rough country, a modern composite design might give the best chance for a safe landing.

Chapter 12
Mold-making

Introduction

A mold is a tool which defines the surface of a part and holds it in place while the part hardens. A mold does not apply a significant force to shape the work. (A die, in contrast, forces the work into a shape. An example of a die is stretching sheet steel to make an automobile fender.) Usually, the surface of a mold is faithfully reproduced in the molded part, which then does not require further finishing or shaping. Common bottles of glass and plastic are made in molds.

The attraction of the moldless construction methods, previously discussed, is that the builder doesn't have to make a mold in order to end up with a finished part. He does have to finish the surface of the part, however, and this might be almost as much work as finishing the surface of a mold.

There are many situations in aircraft construction where using a mold is the only practical option or where using the mold saves work even for one piece. A mold is generally referred to as a *tool* in aircraft and composite-industry practice.

In this chapter, we will consider the general principles of making simple molds/tools. In the next chapter, we will consider more advanced techniques of using molds to make large precision parts.

We will make reference several times to the "prepreg strip" technique described in Chapter 9, so a review may be useful.

The Sheet Mold

The simplest mold is a flat, stable surface of any material stiff enough to keep its shape during handling. The surface of the mold will be reproduced accurately in the molded part. Usually, we want a smooth surface. Sheets of glass, plastic, or metal all work fine. One of the very best, most convenient, and inexpensive is a smooth coated chipboard with the trade name Kortron, sold at building supply stores.

Most current composite aircraft designs require that the builder make up some parts with multiple laminations on a flat surface. Often, for small parts, it is convenient for the builder to use any handy scrap of plywood, tape a smooth piece of polyethylene plastic to it to serve as a release film, then lay up the laminations as required. After cure, the part is easily pulled from the "tool" and trimmed as required.

Many useful things can be done with flat tools. Reference marks can be scribed into the surface of the tool so that these scribe lines appear as very slight ridges on the finished parts. A small nail point can be placed in the surface so that it makes a dent into the molded part for a pilot hole for accurate drilling. A smoothly blended joggle can be glued to the mold so that the finished part will have exactly the required bonding area offset to join with some other composite part. It may be convenient to mold a part on a flat tool even though it may be used later use in a gently curved form. (An example might be a lower wing skin which is nearly flat.)

A variation on the flat sheet mold is a curved sheet which has been formed by bending with external ribs into the desired shape.

The wing skins for Rutan's round-the-world Voyager were made on a bent sheet form. Bending and holding large thin sheets to a very accurate form without waves must have been a great challenge. The Rutan crew applied the sandwich principle by using two layers of sheet metal with epoxy between to get the needed stiffness, but the alignment problem remained!

Forming a Male Plug

For many composite applications, where there are complex double curves, there is no practical alternative to the use of a female mold. To make such a mold it is first necessary to make an accurate copy of the surface that the final part is to have (called a *plug*). Often, this surface is not defined exactly and requires trial and error before the final curved surface is decided. Good examples are engine cowls and fuselages.

We will demonstrate the general procedure by describing how to make a mold for an engine cowl, then extend this to the problem of making a mold for a fuselage section.

Tony Bingilis, in his book *Sportplane Construction Techniques* (revised 1992) gives a thorough, well illustrated, detailed description of how to make a plug for an engine cowl. Essentially, the mounted engine is used to support a rough form of foam and plaster which is shaped in place to the desired contour. Bingelis discusses many real-life details of engine protection, access, exhaust clearance, and installation that are specific to cowls. This book is highly recommended to anyone who must make a custom cowl design, as for a non-standard or automobile engine installation.

The Plaster-Surface Plug

Bingilis gives many useful details on how to locate and carve a foam base to the desired shape. He suggests that the foam be rough carved to a level slightly smaller than the finished dimension of the cowl surface. Next, the level is brought up by coating the foam with a drywall finishing plaster, such as Ready Mix All-Purpose Joint Compound. (Recall that joint compound, as used in the drywall industry, sands out very well.) This soft plaster is brought to final dimensions by sanding. As in finishing a wing, a smudge stick is useful for finding local high spots. After the plaster has been sanded to an excellent surface and has hardened completely (24 hours), the surface should be inspected and sealed with paint.

The Foam and Micro-Surface Plug

An alternative is to finish the foam plug the same way we finish the foam wing in Chapter 9. The method would be to shape the foam as close to the final dimension as is convenient, then fill with dry micro. After the dry micro fill has hardened, low spots are filled in with dry micro and the surface is carefully reshaped, inspected, and filled again. In case of serious mistakes of over-sanding, slices of foam can be added and attached with 5-minute epoxy. (The small hot wire saw and the "baloney slicer" can be handy here.) When the filled surface is very good, it can be painted and used as-is, without a glass skin.

Finishing the Plug

Work will be saved in finishing if a surface filler such as Feather Fill or Morton's Eliminator is used at the final sanding. These materials will fill pores and pinholes and produce a tight smooth surface to hold the separation coating.

If the finished part should have any reference lines or drill hole marks added, this is the time to do it. A little extra time spent here may save a terrible mistake later. Use a countersink bit in a hand drill to very carefully place cone-shaped dimples wherever any pilot holes should be. These dimples will produce high spots on the mold which will make accurately placed indentations in the finished part.

Reference lines are probably best determined while the plug is on the aircraft. A possible small problem is that scribing a scratch line on the plug produces a raised line on the mold and so makes a groove on the finished work. Scribing the mold produces a ridge on the part, but it is often harder to locate the desired line. If it doesn't matter, do it the easy way and scribe the plug. For recovery in case of errors, it is a good idea to place reference holes on the plug wherever they might be needed and decide later whether to fix the holes on the mold or on the work.

After adding reference scribe marks, the surface should then be painted with any spray can of gloss lacquer paint that might be available. Inspect the work. Smooth and repaint any re-done areas.

Mark the plug with a black "permanent" laundry marker wherever it will be necessary to divide the female mold into sections.

Large Plugs

Large molds require large plugs. To make a mold, we must first have a stiff, cheap, fast, and exact replica of the outside surface of the major parts of the aircraft. The author has not verified it, but has heard in the industry that Burt Rutan came upon his basic moldless techniques while looking for a good way to build only a plug. The "plug" turned out to be so light and strong that it became the actual structure!

The builder may wish to provide for commercially available components, such as a ready-made windshield or salvage landing gear parts, which can be blended into the plug shape.

There is no single best method of making a plug. Any method is a candidate which lends itself to accurate carving and which will support a reasonably strong surface. (This is one reason why Rutan's hot wire foam cutting is so attractive.) Nearly all large plugs are built over some kind of steel skeleton as a base support. The method most often mentioned in the literature is to use a wire mesh over plywood formers and stringers, which is then covered with plaster of Paris. The resulting plug is heavy and hard to finish.

More recently, fuselage plugs have been made by first making a rough steel skeleton, then attaching large blocks of styrofoam to the skeleton. The styrofoam is carved to shape by hot-wire cutting, bread knives, power sanders, or whatever, and given a smooth surface that can stand reasonable handling. Aggressive shaping with such brutal machines as power belt sanders is acceptable in the interest of speed because repair is easy.

As referenced previously, Bingilis recommends surfacing foam with joint compound. Hollman recommends a single skin of glass, finished with Bondo. (Remember that Bondo dissolves styrofoam, so the epoxy/glass skin must be in place before Bondo can be used.) Three-ounce crowfoot can be used as the skin layer. If the builder chooses to go through the glass skin, it is easily replaced and the new fine-textured surface is easily filled. Automotive fillers like Feather Fill or Sterling Primer/Filler can be used to develop the final surface.

An important source of plugs is another existing airplane. A number of successful designs, including the *Dragonfly*, have been offered as prefabricated kits where the kit manufacturer has used the original aircraft to shape the tooling molds. In the case of the *Dragonfly*, the molds were made in hemp-reinforced plaster in a few days by Task Research. The plaster molds were faced with fiberglass for a finish surface. (For the *Dragonfly* kit, the decision was made to make the wings by the Rutan method, as originally designed.)

Molding Directly on a Plug

Bingelis makes the assumption that the builder will choose to make a layup of several layers of fiberglass on the mold and use that layup directly as the finished part. With this scheme, a female mold is not used at all. The builder must then finish the outside surface of the molded shell.

The hardened shell is cut into sections, if necessary, with a hacksaw or Dremel saw and is then removed from the plug. (The shell will be thin, so this will be easy to do.)

If the builder is somewhat lucky, the finished mold shell can be removed from the plug in one piece by pulling it forward. An engine cowl can be a particularly difficult case because of the large horizontal areas and the reentrant curves around the cooling air intakes. Recognize that it will probably be necessary to divide the mold into sections to get it off the plug. The nose bowl, for example, may need to come off, requiring that the shell be repaired with a butt splice. (Realignment of the mold sections will be easier if the cut line is made with about three "S" wiggles in it.)

There is some economy of effort with this method, even if the shell must be cut, because the surface finish of the plug needs to be only good enough to get a reliable release of the layup from the mold. The shell will be larger than the plug by the thickness of the shell. If the plug is attached to the aircraft, the method requires that the builder work on an open surface with wet resin on the under side of the cowl. Resin is not very thixotropic, so will slowly ooze toward the low point until hardening begins, making the bottom of the shell resin-heavy.

Making a Female Mold

Better practice would be to make a female mold and lay up the cowl in that mold. This requires that the plug be very well finished because the finish will be carried into the female mold. The female mold will then produce a part which will require very little work to finish. The female mold also makes it easy to include surface stiffeners, joggles for access doors, or recesses for latches.

An important benefit of making a female mold is that most of the work has been done when the mold exists. Additional parts can be produced with very little additional work. For the case of a straight-winged airplane, one wing mold can be used for both the left and right wings.

Making a Separation Line On the Plug

We will usually prefer to make an accurate partition line in a mold, rather than just lay up a one-piece shell and cut it into sections. To do this, attach a temporary flange of any smooth rigid sheet material (like aluminum) to the plug, using an adhesive which can be broken free and sanded out easily. A filler like Superfil Epoxy Filler works well. Attach the flange on ONE SIDE ONLY, because we will lay up a section of the mold against the other side.

Finding the accurate trim line for a close-fitting partition flange can be awkward, but the following technique works:

Stick a thin strip of wood, like 1/8 inch basswood purchased at hobby shops, to the desired line with tiny patches of Bondo. Use duct tape to hold the strip in place until the Bondo sets. Remove the tape. Cut a piece of plywood to correspond very roughly to the shape of the line and hold the plywood next to the basswood strip with duct tape. Bondo the plywood to the basswood, either directly or via little scraps of wood. When the Bondo hardens, we have our accurate line. Break off the Bondo patches under the basswood strip and sand off the Bondo residue.

For extreme curves, like around the front of a cowl, use a piece of #12 electrical copper wire instead of a wood strip. Five-minute epoxy is better than Bondo for attaching to the bent copper wire.

Linoleum and tile shops sell a useful tool for trimming tile, which consists of a great many short parallel pieces of steel wire held under two straps. The tool looks like two combs back-to-back. This tool is handy for finding and transferring short-radius curves.

Making the Partition Flange

When the line of the partition is found, transfer the line to the partition flange and cut it to shape. One very good way to make a partition flange is to mark the line with a "Sharpie" pen on a sheet of polyethylene plastic, turn the plastic over so that the mark is on the outside, and make up a two-ply prepreg strip from scrap glass. Cut the wet prepreg on the line, trim the outside to make the width about two inches, and let the strip harden. It is convenient to make all of the separation strips that you will need during a single session - this will save time and material. The resulting strips will fit accurately and will provide a smooth release surface for the layup.

Choose a section of the mold that is to be laid up first and bond the partitions to the plug with SuperFil as needed, being careful to keep the bonding on the side AWAY from the layup.

It is a good idea to mark the location of the separation line into the plug with a scribe line. The plug is the only absolute reference we have and we will wish to avoid surprises if changes are required.

Note that additional separation lines may be required to provide for overlap areas for joggle joints. For example, if the fuselage tail cone is to be assembled with two large moldings, say top and bottom, and if a generous 1 1/2 inch bonding area is to be allowed, an overlap of 1 1/2 inches must be provided. Another example is a full-span wing leading edge molding which must overlap the top and bottom wing skins to close the wing. The appropriate joggle offset must then be added to one or both molds which will be taken from the plug. This requires that, after one mold has been taken, the partition must be knocked off of the plug, the broken filler cleaned up, and the partition moved to the new offset location. (This is where accurately marking the old location of the partition is important!)

Adding a Bonding Area

Usually, bonding areas for joints can be made as described above. The best practice is to provide for a large-area self-aligning joggle joint, even though very good butt joints can be made. If it is necessary to add an area to the mold shell along the edge where we wish to have the joint, the following procedure will work:

Remember that only the face of the mold carries an accurate surface. We have built up the mold shell in a rather casual way by stacking up strips of wet prepreg and accepting whatever thickness resulted. We need to extend accurately the line of the face of the mold to have a reference for the joggle.

Make up a batch of short pieces of hardened 2-ply glass laminate prepreg about one inch wide and four inches long. Let's call these short pieces "extenders". Tack Bondo about two inches to the mold surface and leave about two inches hanging over the edge. Use as many extenders as required to define a smooth curve.

Prepare a strip of hardened 2-ply prepreg as long as the edge that is to be extended. Call it the "joggle support." It may be necessary to fit the joggle support to the edge by the same method used to obtain a line for the partition flanges described above.

Clamp the joggle support to the line of extenders so that the surface of the joggle support is aligned with the surface of the mold. Use wet prepreg strips to firmly attach the joggle support to the edge of the mold. Let the work cure. Remove the clamps.

Use a wood block and a hammer to knock off the Bondoed extenders. Clean up the Bondo and fill any gaps between the mold and the joggle support with dry micro. Finish sand the joint. The mold surface now has a smooth extension.

Laying Up the Female Mold

We will make what is called a "splash" mold in the composite industry and will make excellent use of whatever scrap glass and left over glass trimmings may be in the shop.

Note that general industry practice, including kit manufacturers, is to use a special gel coat as the first coat when making a tooling mold. This is to produce a hard surface, extend tool life, and provide a color that makes inspection of the surface easier. It is not essential to use a gel coat and the homebuilt aircraft suppliers do not stock the material. The following description will assume that no gel coat will be used, where a procedure for the homebuilder is discussed.

We will want a strong sharp edge where the mold surface meets the partition, so mix up a little flox and make a 3/8 inch radius along the partition. The human index finger is the perfect tool. (Mix microballoons about equal to the dry flox for better handling.)

The plug should be well waxed with general-purpose Carnauba floor wax. Three coats, well rubbed, should be sufficient to assure parting. Commercial water soluble mold release agents are sold by the supply houses and may be preferable because they are less likely to contaminate joints with wax. Most sources recommend the use of both wax and PVA mold release.

Refer to Chapter 9, where we describe how to make "prepreg" strips. Use epoxy, rather than cheap polyester, to avoid the shrinkage associated with polyester. Make up a generous supply of 2-ply "prepreg" strips about 2 inches wide and 12-14 inches long, depending on what scrap glass is available. Squeegee them out, but keep them a little wet. Fiber orientation is not important, but if bias cut scraps are available, use them that way. If you want to be elegant, stack a ply of bias cut with a ply of straight cut. Cover the plug with overlapping prepreg strips, criss-crossing the plug so that you end up with about four layers of glass, on the average, everywhere on the plug. Stipple out any air and strive for contact everywhere with the plug surface. This operation will take only a few minutes and the dripless prepregs will be particularly appreciated when working underneath the plug. If the viscosity of the resin is unusually low so that the second layer of prepreg strips doesn't want to stick to an overhead surface, don't worry about it - use one layer. The basic purpose of this step is to capture the shape of the plug surface. As long as there are no voids, the operation is a success. Tidy up after the job with a knife trim, if convenient. Let the "splash" layup cure. If larger pieces of glass are conveniently available, use them, wetting out directly on the work.

After the first layer of glass has cured, inspect the new mold shell. Add more prepreg strips wherever the glass appears to be thin, particularly in areas of high curvature, but this is not critical.

After the layup has hardened, remove the partition flanges and clean up the broken SuperFil.

Apply mold release to the next section of the plug, including the smooth vertical surface produced by the partition flange, just removed.

As before, make a flox radius at the partition, lay up the next section of the mold, knife trim when ready, and let the assembly harden.

We now have a very accurate reproduction of the plug, divided into sections.

The next task is to assure that the mold holds its proper shape when removed from the plug and can be supported conveniently while being used to produce the final molded part. There are two choices: (1) the fast, cheap, one-shot, and (2) general industry practice.

The Fast, Cheap, Mold

This method saves resin and time and is based on providing minimum support for the simple "splash" shell described above, generally assuming that the builder wishes to build only one part.

Cut scraps of plywood or scrap sandwich stock into strips about two inches wide and set them edgewise as stiffening spars wherever there is a large unsupported area. Tack the stiffening strips into place with Bondo. (This is a good re-use for the partitioning flanges!) Secure the stiffeners with wet prepreg strips laid along the corner intersection. The idea is to form an external stiffening cradle and support for all sections of the mold after they are removed from the plug. Let the ugly mess cure and remove the mold sections from the plug.

Use scrap wood, Bondo, and screws to rig legs to support the mold on the floor or work bench.

Industry Method (Permanent Tooling)

Industry practice is to add thickness to the skin of the female mold to make a rigid structure without local area stiffeners. Weight is not a significant factor in a tool, so the skin is made thick by adding a layer of matting (unwoven glass felt), then a final layer of glass cloth for strength and to tie down the sharp glass fiber ends. Mat is often seen in boat construction, where it is a cheap, easy way to build up core thickness. It is heavy because it serves essentially as a resin-filled spacer between glass skins, so is very wasteful of resin. An alternative is for the homebuilder to use extra layers of glass to build up thickness, using out-of-date epoxy and scrap (old) glass which he would not use in a flying structure.

Support pads are added as a part of the reinforcing layup to provide attachment points for a steel frame which supports the entire mold. Adding caster wheels to the frame will make it easier to use the finished mold.

Note that an additional factor in mold design for the commercial industry, including the prepreg kit builders, is that tools of this type will usually be used with high-temperature cure prepreg materials, hence will be used in an oven. Care is taken in the design of the supporting steel frames to allow for heat expansion, including the use of slotted holes for mounting bolts. Special tooling resins are also used, which are designed for high temperature service and low expansion. For very high performance military and air transport applications, where very high temperature cures are used, all-carbon tools are used to produce all-carbon parts.

Preparing the Female Mold

For this discussion, we will assume that the builder has completed making a "splash" mold shell and has repaired the shell as necessary after any cuts. Clean off all the mold release material with whatever solvent is necessary. Wax will be much harder to remove than water soluble releases.

If the molded surface is to have any recesses, like vent intakes, latches, or joint joggles, they should be added at this stage. Sliced foam can be sanded to the shape of the recess, bonded to the shell with 5-minute epoxy, and faired in smoothly with dry micro. A skin of 3-ounce crowfoot glass will protect the foam surface. Bear in mind that sharp radius curves in the glass layup will not be practical. If a sharp radius is required, it will have to be added later with a flox corner.

Making the Joggle Offset On the Mold

If the molds that defined the several parts of a large assembly like a fuselage were perfect, the parts made from them would meet exactly at the partition lines of the molds. A "perfect" joggle joint would require a 90° right angle in the fibers going into the joggle area. Even if this were practical, the stress concentration associated with this turn would be unacceptable. The edge of the joint must be tapered, as shown in the sketch (repeated from Chapter 7). The theoretically perfect edge of the molded part is cut back slightly to allow for the taper and a joggle offset is provided to accept the other part to be bonded. The small gap is filled when the aircraft is finished.

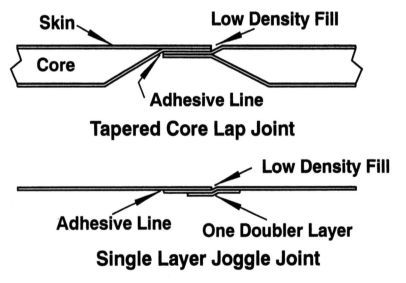

If the joggle offset only needs to accommodate around 0.060 inch, typical of many skins, the offset can be made the easy way by sticking release tape of the proper thickness to the mold as required. Blending the joggle offset is generally required only for offsets of 0.1 inch or larger.

If a large joggle offset is required, use a stack of prepreg strips on the mold to duplicate the part that will be attached. Smooth the transition with dry micro. Let everything cure.

Chapter 13
Using Molds

General Remarks

In this chapter, we will discuss the techniques of making parts from molds. One benefit of using a mold is that quite complex shapes can be produced with little additional work, no fasteners, and requiring very little finishing. Materials can be used efficiently to achieve high strength/weight performance. Additional parts can be easily made. For example, a group of three builders might work together to produce a single set of molds which define an aircraft, then make three copies from the single set of molds.

This chapter will describe the process of vacuum bagging in some detail. We will show that any builder who would feel confident undertaking to build a Rutan-type airplane could easily use the process to make more complex parts. It has been the experience of the author that builders who have actually seen the process demonstrated come away with confidence that they can use it themselves.

Basically, the molding methods to be described produce precision sheets of very strong material. The molded sheets can be flat or can have complex curves almost without limit. A single molding operation can include structures, such as sandwich panels and/or stringers, to provide local stiffness wherever desired. A great advantage of the technology is that large structures like entire wings or fuselages can be assembled from only a few of these accurately formed sheets, with a minimum of internal structure. Compared to earlier frame-and-fabric or metal technologies, the contrast in parts count is quite striking.

Individual molded parts may be simple, but the detailed design of complete structures requires a qualified designer. The designer has many options and many tradeoffs to consider in making design decisions. Among these are:

1) Distribution of skin thickness, fiber type, and fiber orientation, considering bending, torsion, and handling loads,

2) Size, depth, and span of stiffener stringers and/or sandwich panels,

3) Design of spars, including caps and shear webs, and their means of bonding,

This chapter will describe, in general terms, examples of three typical molding operations. These examples are only to demonstrate procedures are not to be taken as descriptions of an optimum structure.

Making an Open Layup In A Female Mold

Before making any layup, inspect and clean the mold carefully and treat the surface generously with mold release. Remember that every speck of dust and every dead bug will be faithfully reproduced, so remove them first.

Unlike the happy-go-lucky process of making a splash mold, we are now making a structural part and will follow the designer's layup schedule. The designer will specify how much glass or carbon goes where and in what order.

Because we are using a female mold, we have an opportunity to save weight because we can use a thin skin with stiffening only where needed. The concept is illustrated in metal by the engine cover door used on the *Bonanza*. The *Bonanza* uses a curved stamping of thin sheet aluminum with a corresponding internal hollow hat section stiffener of stamped aluminum. The two pieces are held together with many rivets to make a combination stiff enough to be useful. This method can be duplicated in composite by using a female mold and adding stiffeners to the inside of the skin.

The simplest case is to lay glass into the mold to make a skin of uniform thickness without added stiffeners. The process is the same as described in Chapter 9. (If a bond to another part is required, peel ply must be laid over the bond area before beginning the layup.) To minimize overlaps, the builder should use pieces of glass which are as large as convenient. Working in a female mold is a bit harder than working on an open or convex surface and smaller pieces of cloth will be needed to fit into sharply curved areas, such as a nose bowl. The prepreg strip technique is useful to place cloth, particularly in sharply curved areas. The plastic backing of the prepreg strip cannot drape over a dished curve, but the wet cloth can be brought very near to the correct position and formed after the plastic is removed. The layup is complete when the specified thickness is reached. The builder waits for the knife trim, tidies up, waits for the cure, and the job is done with very little finishing required. Most fiberglass cowls are built this way.

Vacuum Bagging - Introduction

Many of the readers of this book will likely be active members of local chapters of the Experimental Aircraft Association. As pilots and builders, they will have seen the work of leaders in the field, excellent workers with great skill and knowledge who are ready to share their skills with their friends. Since the great success of the Rutan family of aircraft, there is hardly an EAA chapter that does not have at least one Rutan-type aircraft flying. Even those who prefer the older technologies are willing to use "fiberglass" for cowls and fairings and will cheerfully call upon their "composite" friends for aid and advice on how to use "plastic" to do these things. At the mention of "vacuum bagging", however, there is a general run for the exits. "Vacuum bagging" is seen by many builders as being the proper province of Boeing, NASA, the very rich, and the best kit manufacturers.

Vacuum bagging is the process by which the several parts of a layup are forced into very tight contact, using atmospheric pressure, which is as high as 2,000 lb/ft^2 with an ordinary vacuum. This greatly reduces the risk of voids or poor interlamination bonding and removes the excess resin which is typical of wet layups made by hand. The parts produced by the homebuilder can approach the performance of those made by fully-tooled commercial shops using industrial prepregs. The amateur will not have the commercial prepreg and the high temperature curing ovens to get the very best properties that the kit manufacturers can reach, but wet layup, vacuum bagging, and a backyard post-cure at 140°F give a very satisfactory result.

The aircraft supply houses offer the special supplies required for vacuum bagging operations in the small quantities needed by the individual builder. In general, these are industrial supplies intended for high temperatures and high pressures which are not used by the homebuilder. Ordinary materials from the hardware store can often be substituted and these substitutions will be discussed.

Summary of the Basic Vacuum Bagging Process

The sketch on the following page shows the relationship of the components that will be discussed in detail below.

The Tool

The tool (or form) supports the work in the shape that is required. The tool can be a smooth flat plate, a deeply curved cowl form, or a large form for a wing skin or a spar. The tool needs to be strong enough for ordinary handling, but does not need to be designed to support large forces, like a metal forge or a steel press. As long as the tool does not leak air, large forces are placed across the tool and the work, but the forces on the system balance to zero. Consider that a scuba diver at a depth of about 100 feet has no feeling at all that the balanced pressure on his body is over three tons per square foot; breathing air from a regulator at exactly ambient pressure, the air in his body cavities is at the same pressure and he feels nothing. A deep-diving animal like a whale or seal is designed to have the air in its body cavities compressed to a very small volume (high pressure), so the animal also feels no pressure. A whale would understand vacuum bagging.

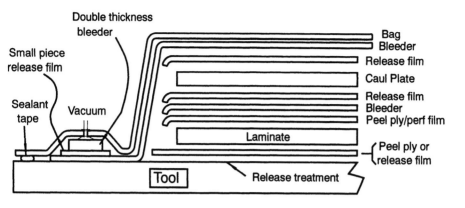

Typical Vacuum Bagging Stack

The Vacuum Chamber

The vacuum chamber is formed by a "bag" of very soft plastic film which is attached around the periphery of the tool by any convenient means, usually a line of sticky plastic tape sold in the industry as "bagging tape" or sealant tape. The soft plastic bag, of course, cannot carry any load, so is pressed by the atmosphere into hard contact with the tool or anything attached to the tool that the air cannot get behind. The bag, then, can only serve as a seal and must be supported by some other rigid object.

The Air Path - the Bleeder Ply

A means must be provided for air anywhere in the vacuum chamber to find its way to the vacuum outlet, as the soft bag wrinkles and shrinks as it is forced against the tool. This means, called the bleeder, is a porous fabric that will allow air to flow laterally even though the fabric is being compressed. Industrial bleeder material is designed for use in a pressurized autoclave. For the homebuilder working at atmospheric pressure and room temperature, almost any absorbent cloth will work as a bleeder - cheap fabric store quilt stuffing works fine.

The second important function of the bleeder is to absorb excess resin from the laminate.

Suppose we have a rigid flat tool, like a sheet of Kortron, put a circle of bagging tape on the surface, put bleeder material inside the circle, and cover the assembly with soft bagging film, being careful to see that the bagging film is sealed all around the edge. As we pump the air out we see that the bag crinkles up hard against the entire tool surface, leaving no voids. Essentially, the sealed bleeder cloth has provided an exit path for all air in the chamber and the bleeder cloth has become a powerful press.

If the tool were not flat, but had a complex shape, like an engine cowl, the effect would be the same as with a flat tool. The air behind the bleeder and bag presses against the air behind the tool. The bleeder accurately assumes the shape of the tool. The pressures are balanced.

Attaching the Bag

The tool must provide a surface outside the work area for attaching the bag. In the case of a flat mold, this is just the unused area outside the work area. In the case of complex molds, some kind of extension or flange must be added outside the mold surface. Dimensions are not critical; as long as a bleeder channel exists to reach the edge of the work area, the bag will work.

The simplest way to attach the bag to the attachment area is with two-inch brown plastic mailing tape placed so about one inch is over the bagging film and one inch is on the mold surface. This way is fast and cheap, but makes adjustment of the bag a little difficult.

The most versatile way to attach the bag is to use special vacuum bag sealant tape. This is a tacky substance that can be easily torn from the roll and formed with the fingers. It has an appearance and texture similar to glazier's caulking. The user simply runs a line of sealant around the attachment area and presses it down on the tool. The bag is then just pressed against the sticky sealant surface. If adjustment of the bag is needed, the bag is simply lifted off, moved to the new position, and pressed down again. The material is not expensive and can be reused. If the tool surface does not take the sealant tape well, a strip of plastic tape can be placed on the tool as a base. The sealant is then placed on the smooth tape surface.

Remember that the bag is assumed to have no strength at all and must be able to conform perfectly to the work without breaking or pulling away from the edge seal. To assure enough slack in the bag, a two-inch pleat should be taken in the edge about every 24 inches. This is a very obvious operation. The builder presses the transparent bag material against the tape, checking to see that no creases exist to cause leaks. At each pleat, the builder tears off enough sealant tape to fill the pleat, inserts the tape, and squeezes the pleat to seal it. This is a very quick process, because everything can be done without trimming the bagging film and any adjustments are easily made. Most leaks are obvious by inspection. Placing a bag usually takes only a few minutes.

The Laminate

The laminate is the stack of layers of fibers and reinforcements that the designer has specified to make up the finished part. A wing skin might consist of a very large area of very thin fibers with foam sandwich stiffeners placed at intervals over the tool. A wing spar might use many layers of glass or carbon in a thick section. Several separate laminates, like a batch of wing ribs, may be run on a flat tool at the same time with a single bagging operation.

Peel Ply

As with the open layups we have previously discussed, peel ply provides a removable surface which leaves a clean, fresh, textured face for bonding. Peel ply is used when vacuum bagging to prevent the bleeder from bonding to the laminate.

Peel ply is manufactured in nylon or polyester. Two-ounce nylon gives better conformability (nylon is stretchy) and a smooth surface finish, but is not recommended for resin systems containing phenolic. Three-ounce polyester produces a rougher surface and is preferred to protect a bonding surface. It does not conform well. (This problem is avoided by using the material in narrow strips.)

In the simplest possible vacuum bagging procedure, the laminate would be covered with peel ply, then bleeder, then covered with the bag, pumped down, and allowed to harden. After hardening, the resin-soaked bleeder is removed with the peel ply.

Usually, peel ply is used to cover the entire surface of the work on both the tool side and the bag side. It is not used at the tool if the designer wants to accurately reproduce the surface of the tool. Peel ply is always used on the tool side wherever the work requires a bond site on the tool side. Peel ply is also useful because it protects the surface of the work from damage and contaminants until it is required for some other assembly. If wax was used as a mold release, the peel ply layer will prevent the wax from contaminating the tool-side surface.

Release Sheet

A release film or release sheet is any material which resin does not stick to. Cheap polyethylene film from a paint store, brown plastic package mailing tape, Saran wrap, and kitchen garbage bags are all good release films which work well at the ordinary temperatures the homebuilder will use. Release films separate freely from the resin surface, leaving a glassy smooth surface unsuitable for bonding, but requiring very little finishing. Industrial high temperature release films cost about the same per yard as structural fiberglass, so the amateur will usually prefer to use the low temperature substitutions available from the hardware store.

Perforated Release Sheet

For the simple case discussed above, where the work (with peel ply) contacts the bleeder directly, the resin extraction process can be too efficient. Under pressure, the resin flows through the peel ply, then flow directly into the bleeder, leaving the work starved for resin.

Perforated release sheet serves as a limiting barrier to resin flow which allows just enough resin to flow through to the bleeder without starving the work. After curing, the release sheet pulls off easily.

Industrial perforated release sheet is available for the large user in a wide range of effective porosities (different hole intervals). The homebuilt aircraft suppliers offer a limited range of industrial types. The builder can make his own, however, by perforating ordinary release sheet with a nail board. Bagging film, or 2- or 3-mil hardware store polyethylene can be perforated to make perforated release film.

The Nail Board

A nail board is very useful for making perforated release film and for preparing foam sheet for bonding in a layup. The nail board is a very useful tool in the home composite shop.

To make a nail board, cut two pieces of 5/8 inch plywood to about 6 inches by 14 inches. In one of the boards, drill a 5 inch by 12 inch array of 1/16 inch holes on 7/8 inch centers. (A drill press will be useful for this many holes.) Insert a 1 1/2 inch 4-penny box nail in each hole and push the nails through until the heads are flush. Glue the remaining board over the nail heads to hold the array together.

To use the nail board to make perforated release film, place several layers of the film to be perforated on a stack of three or four old towels. Place the film on the stack and cover with another old towel. (This towel helps get the film off of the nails.) Push the nail board through the sandwich of towels and film as many times as necessary to make a rough hole pattern over the surface of the film.

Mold Release

We discussed the problem of mold release briefly in Chapter 12. The same mold release materials can be used with bagging on a tool. Some simplifications are possible by using release films directly on the tool face.

Many wet layups are made directly on a piece of mold release film before placing the work on the tool. The whole layup, including the mold release film, is placed on the tool as a unit and smoothed into position. (An exception might be a layup which has unusual curvature or thickness, requiring that additional layers of wet glass be added to the laminate after placement in the mold.)

With a flat plate tool, it is possible to simply tape a piece of mold release film over the tool where the work will touch - no wax or other contaminating material is necessary. With small parts, the pressure of bagging drives out the air between the work and the tool, producing a smooth surface. With large parts, it is theoretically possible to trap a small volume of air between the film and the tool and produce a deformed part. If there is any doubt about a substitute tool release film, use industrial bagging film. If the tool surface is waxed, the tool surface becomes sticky enough to hold the release film and the film can be squeegeed out to remove air bubbles and inspected to be sure that it is free of bubbles. Cheap polyethylene often works fine as a release film. Polyethylene is attractive because it is smooth-surfaced, flexible, and very cheap. It does tend to have creases from packing, however. A hair dryer greatly reduces the creases, but doesn't quite eliminate them. The builder can often work around the bad creases.

The Vacuum System

The vacuum connection must be set up so that it always has free access to the bleeder layers. The bleeder material must always have enough volume to provide an air path even though the portion of the bleeder closest to the work absorbs the excess resin. The builder must not permit the condition where the bleeder fills completely with resin and is no longer an effective air path, but becomes a resin path. We want to assure that the vacuum connection is the very last thing in the entire rig to fill with resin. To do this, the builder places a six-inch square of bagging film on top of the bleeder, folds a double or triple thickness square pad of bleeder on the bagging film, then places the vacuum connection against this extra pad of bleeder.

The vacuum connection is made through a small X-shaped cut in the bagging film. It is most convenient and most reliable to use the same standard high-temperature two-piece vacuum valves used in the industry. These valves are expensive, but can be endlessly reused. At least two are needed for each bagging operation. (The clever builder may be tempted to try to avoid their use with home-made hardware, but the experimental energies may be better applied elsewhere.)

The tubing necessary to connect to the vacuum source can be ordinary 1/4 inch clear plastic as sold in hardware stores and the fittings can be found in the plumbing department.

At least one vacuum gauge is necessary at the work. The vacuum gauge should be mounted in a separate vacuum valve at the work. In a pinch, the vacuum gauge can be connected in the line very close to one of the two vacuum valves. Best practice is to use a second gauge at the vacuum pump. If there is a difference in vacuum readings, the builder should find out why.

The vacuum pump need not be of large capacity, but should be the genuine article. An ordinary vacuum cleaner can generate a useful vacuum, but a vacuum cleaner depends on air flow for cooling, so will soon fail in this service. A refrigerator compressor depends on oil in the refrigerant for lubrication. The suppliers sell a small legitimate oilless vacuum pump for less than $500. A single pump could build a lot of airplanes.

A vacuum regulator is essentially a controlled vacuum leak designed to adjust the vacuum to a desired level. The unregulated pump reaches nearly a true one-atmosphere vacuum, which may produce too much pressure for some layups. For the materials and methods described here, a vacuum of about half an atmosphere produces good results. The regulator mechanism is a tee fitting in the vacuum tube with a small ball bearing held against a seat with a small spring and vented to air. The vacuum is set by adjusting the spring force.

The Caul Plate

The above procedure works well for amateur use. Supplier catalogs and the literature from industrial sources show a "caul plate" in most diagrams of layups. The homebuilder will rarely use a caul plate, but it should be understood.

The caul plate may be used to improve the distribution of resin during the bagging operation. Its function can be understood by considering the simple flat plate tool described above. Suppose a glob of stiff bread dough had been placed in the above bagging setup, covered with bleeder and a bag and pumped down. See that the bag would wrap around the glob of dough in whatever shape matched the glob; there would be a limited internal pressure difference to make the dough flow to conform to the flat tool.

If we placed a flat steel plate (the caul plate) on top of the dough, with bleeder on both sides, see that air would be drawn from between the caul plate and the work. The plate becomes a powerful flat press, moving to shape the dough into a flat form. High spots would receive a concentration of force. If a caul plate had been used in the previous example, areas bulged up slightly from being resin-rich would be forced toward lower resin-lean areas.

A formed caul plate can also be used to shape a part.

Chapter 14
Examples of
Vacuum Bagging Procedures

General Remarks

This chapter will describe, in some detail, typical procedures for making molded parts on a formed tool using vacuum bagging. The purpose is to show the reader the general principles involved, not to give exact instructions for producing actual parts to be flown on a specific aircraft.

The three examples given demonstrate a range of techniques, including different use of release film, different ways of handling reference marks, and methods of laying up large area parts. An engineer developing a design using these techniques would find it necessary to make a number of experimental layups to be sure that the many variables of molding, including perf film, vacuum pressure, clearances and tolerances were correctly specified. The examples will suggest further experiments to the builder.

Basic Vacuum Bagging Operation -
Example 1 The Flat Sandwich

Discussion

We will describe the layup of a rectangular sandwich laminate on a flat tool. This operation will produce flat "sandwich stock" which has many applications. This example might be typical of ribs. Please refer to the sketch.

Procedure - Example 1

1) Obtain a convenient smooth tool surface. (Kortron is excellent.) Apply three coats of release wax to the working surface, well dried and rubbed.

2) Use the nail board to perforate a sheet of 1/4 inch 3 lb/ft2 PVC foam. Scarf the edges of the foam to at about a 45o angle. The angle is not critical, but be sure the edge is sharp. (We wish to avoid even a small vertical transition which must be bridged by the fabric.)

3) Use a permanent felt pen to mark the work area on release film. Turn the film over so that the marks are away from the work. Place the release film over the work area. (Tape it in place, if convenient.)

4) Place peel ply over the work area and wet it in place.

5) Wet one layer of 9-ounce bi-directional glass and place it on the peel ply.

6) Paint both sides of the perforated foam with resin and place the foam on the peel ply

7) Place a second layer of 9-ounce bi-directional glass on the foam and wet it out.

8) Place a top layer of peel ply on the stack and wet it enough to hold it. (This layer will receive excess resin from the structure below, so there is no point in adding much new resin.)

9) Place a sheet of perforated release film over the stack.

10) Press a line of sealant tape on the tool so that it makes an enclosure around the work area. Allow extra room at the side for the vacuum connection.

11) Place bleeder cloth over the entire work area with enough added margin to cover the work and reach almost to the sealant tape. Add enough extra bleeder to reach the vacuum connection

beside the stack. (We don't want the vacuum connection fitting to bear directly against the work.)

12) Locate two separate places at the side of the work area for the two vacuum valves. Put a six inch square scrap of release film on the bleeder at each location. Make two pads of double-thickness bleeder and put one on each scrap of release film. Separate the halves of the vacuum valves and place the bottom of the vacuum valve at each location.

13) Cut a piece of bagging film so that it fits quite loosely over the entire tool. Press the edge down over the sealant tape, but leave about a two inch pleat every 1 1/2 feet. Fill each pleat with a short piece of sealant torn off of the roll. Inspect the bag attachment and press out any creases or folds that could leak.

14) Make an X-shaped cut about 3/4 inch long at each vacuum valve location. Press the bottom of the vacuum valves through the cut and connect the tops of the vacuum valves.

15) Connect the vacuum tubing and the vacuum gauge to the vacuum valves and run the tube to the vacuum pump. Turn on the pump and listen for leaks as the bag pumps down. Find and fix the leaks as needed. Work out kinks as the bag finds a stable position. Verify the vacuum pressure. Continue with vacuum for the cure period.

16) After cure, turn off the pump and unbag the work. Remove the work from the tool. Discard the perforated release film and the used bleeder material. Keep the peel ply on the work to protect the surface.

The above procedure produces a tightly-bonded light weight plate of sandwich stock. This material can be cut on a bandsaw to form ribs.

Note that the uncut sheet is quite stiff for its weight, even though the skins are a single layer of cloth. Resin flowing through the nail holes in the perforated foam core makes a characteristic pattern which can be seen through the skin and verifies that the skin has bonded well to the foam.

Vacuum Bagging - Example 2
Long Leading Edge Skin

Discussion

For this example, we will assume that a mold has been made, as described in Chapter 12, from a plug that exactly represents the desired shape of a wing. We wish to mold an accurately shaped skin section that can be used to close the gap between two other wing skin moldings. This skin section must fit exactly over a joggle area in the other two wing skins. We will also assume that the original plug has been carefully scribed to mark the trim line that exactly meets the joggle in the matching skins.

For ease of construction, it is not necessary that this mold have the usual partition flanges described in Chapter 12; it is only necessary that the mold extend beyond the trim line far enough to allow for a line of sealant tape to hold the bag. See the sketch on the following page. It is also very convenient to make two strips for the joggle offset at the same time we make the main skin. These strips will be about 1/4 inch wider than the actual joint to allow for trimming. This procedure gives us a joggle which exactly matches the skin it is to join. The ends of the mold will require extra area for sealing.

Essentially, the mold is a long trough which is sufficiently extended on all sides to allow for sealing the bag. In this case, we choose to make the joggle strips as part of the same layup, so extra area must be allowed. Except for the curved shape and the considerable length, this mold works the same as the flat tool described above.

Generously sized reference marks (indentations) copied from the original plug should be placed around the area of the finished leading edge so that the trim lines can be located accurately. Use two or three marks along the length of the skin so that there is no doubt at all about the boundaries. (After the work is hardened and removed from the mold, we will drill out these marks and use them to snap a chalk line to find the trim line. It is very much easier to repair a small mark we do not need than to try to accurately locate a point on a smooth curved surface without a reference!)

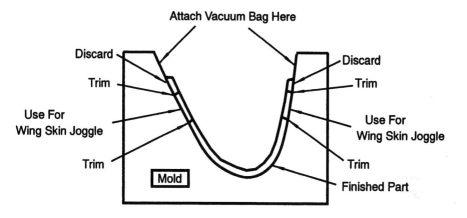

Leading Edge Layup With Joggle

(Skin Thickness Greatly Exaggerated)
(Bagging Materials Not Shown)

Procedure - Example 2

1) Apply mold release, as described.

2) Cut sufficient pieces of 9-ounce bi-directional cloth to make three layers. Allow the width of the leading edge skin plus the width of two joggle joints, plus 1 1/2 inches for trim. Cut two layers of peel ply to correspond.

3) We will use the "prepreg" method to handle the wet laminate. Mark a piece of bagging film about four feet long (longer is too hard to handle) with approximately the outline of the cut cloth. Turn the marked film so the marking is away from the work surface.

4) Place peel ply on the marked bagging film, wet it out enough to stick, and smooth it. Place one layer of glass over the peel ply, pour resin over it and squeegee the resin enough to very approximately cover the area.

5) Use a layer of cheap 3-mil polyethylene to cover the wet glass. Squeegee out the resin and chase it around with the squeegee to be sure that the peel ply and glass is everywhere wet through. (With a little practice, the builder learns to judge the amount of resin required so that very little is wasted. Excess resin will be removed by the vacuum bagging operation.) Rough trim the sandwich of film and wet fabric, leaving a margin.

6) Place the wet prepreg in the mold, starting at one end of the long mold, with the bagging film toward the mold and position it carefully. Use the squeegee to smooth the laminate against the mold face and work out all bubbles. Remove the polyethylene film backing. (A hair dryer may be used to speed removal of the polyethylene, though this is rarely necessary.)

7) Repeat steps (3) through (6) with additional pieces of bagging film, peel ply, and glass to complete the first layer in the mold. Carefully butt each new piece to the previous piece so that they are very close but there is no overlap.

8) Repeat steps (3) through (6) to make enough prepregs to complete the remaining layers of glass in the mold. For the second layer, it is not necessary to use bagging film; polyethylene film is perfectly satisfactory as a throw-away prepreg carrier. Do NOT include peel ply on the second layer. DO put peel ply on TOP of the third layer. Squeegee out the resin and trim.

9) Remove the polyethylene from one side of the prepreg and position the prepreg in the mold. Locate the new layer so that it bridges the small gap between the previous sections. Remove the polyethylene backing.

10) Continue the process until all three layers of glass are in place with the gaps covered.

11) Place perforated release film over the work. Allow an overlap of about one inch where adjacent pieces meet. Place bleeder cloth over the entire work area, with a patch of release film and extra bleeder at each vacuum valve location. (For a long molding like this one, it would be safe to use four vacuum valves.) Place the

bottoms of the vacuum valves on the bleeder, off the surface of the work.

12) Place sealant tape around the work area and cover the mold with bagging film. Where necessary for this large mold, the bagging film can be spliced with 2-inch brown plastic mailing tape. Allow plenty of slack for the bag to go into the mold freely. Put pleats in the bag every 1 1/2 feet. Connect the vacuum system and fix any leaks.

13) Allow to cure under vacuum. Unbag after cure. Discard the perf sheet, the mold release film, and the resin-soaked bleeder. Leave the peel ply.

14) Locate the reference marks (dimples). Drill 1/16 inch holes and insert nails. Use the nails to hold a chalk line and snap the chalk line to find the trim lines. Go over the chalk marks with a permanent marker pen. Trim on the lines. (A fine-tooth electric jig saw works well. Best is the kind with orbital action.)

We now have a very strong, accurately shaped, leading edge skin and two long strips that exactly replicate the required joggle offset needed to attach the new part.

Vacuum Bagging Example 3
Upper Wing Skin

Discussion

We will assume that a mold has been made from the same plug that produced the leading edge skin above, as described in Chapter 12.

The mold must provide areas outside the work area to attach the vacuum bag. At the leading edge, this area is just the surface of the wing carried to somewhere beyond the joint to the leading edge skin. See the sketch on the following page. The mold should carry reference marks which accurately determine the area of the joggle joint. At the trailing edge, the mold must be provided with an extension area, of any smooth material.

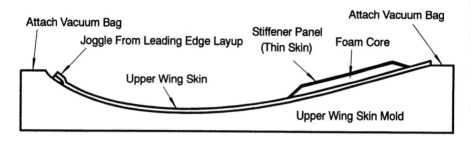

Upper Wing Skin Layup - With Joggle and Stiffener

(Skin Thicknesses Greatly Exaggerated)

Find the reference marks on the mold that locate the leading edge joggle joint and snap a chalk line on the marks. Mark the line with a permanent felt tip pen.

Attach the joggle strip for the UPPER skin from Example 2 to the mold with structural adhesive. This strip produces an offset which is an exact match to the leading edge skin it was cut from and even allows for the thickness of the glue joint.

The mold is a good illustration of the versatility of this method of construction. Whatever skin layup we make, of whatever skin thickness, and however the spar and ribs are designed, will match the wing surface as defined by the original plug. Further, it will attach accurately to the leading edge skin because the joggle matches.

The mold design offers the option of laying up the spar caps at the same time as the skin, though most designs choose to keep them separate.

The designer will recognize that the limiting factor in the structure will usually be failure from buckling in compression and this factor will determine stiffener design. The skin, with whatever local stiffening is provided, shares loads with the spar system in a complex way.

For this example, we will assume that the designer has chosen to make the spar system separate from the skins and will bond the skins and spars during a separate assembly operation. Spars provide convenient attachment points for much internal hardware, like control actuators, landing gear hard points, wiring supports, pitot tubes, etc., and these items are easier to work on before the wing is assembled. The example, then, will outline one procedure for molding only an upper wing skin with associated sheet foam stiffeners.

For simplicity, it will be assumed that a single sheet of triax stitch-bonded fiber (+45°, 0°, -45°) will be used for the main skin.

Procedure - Example 3

1) Use a permanent felt tip marker to mark the dry mold with the location of stiffened areas and ribs. These marks will be visible through the release film and the wet layup.

2) Cut the peel ply sections, the triax glass, and any additional glass layers that will be required for the layup. Roll them, label them, and keep them clean.

3) Use 1/4 inch 3-pound PVC foam for stiffener core. Perforate with the nail board and cut to size each piece of foam. Scarf (chamfer) the edges as specified. Mark which panel goes where.

4) Prepare the mold, as before. Wax the mold very thoroughly. Run a line of bagging tape around the entire work area, but leave the protective tape in place until ready to place the bag. (Messy work ahead!)

5) Lay mold release film (bagging film) over the work area. Smooth out the release film so that it sticks to the mold release wax. For this large area, it will not be possible to avoid gaps in the release film without overlapping, so butt adjacent pieces as accurately as is reasonable. Trim.

6) Lay peel ply on the leading edge joint area and wet to hold in place.

7) Place the cut triax glass over the work area and wet it out, working from the middle outward. As the glass wets, the marks on the mold that locate the stiffener panels will show through.

8) Paint both sides of each stiffener panel with resin and place it over the corresponding marked area. The curve of the wing section will make a slight gap between the curved wing and the straight foam, but the wet triax will hold it sufficiently. The designer will have allowed plenty of clearance between the edge of a stiffener and the nearest rib or other structure. Recall that it is not necessary that the stiffened panel be continuous.

9) Place 3-ounce crowfoot glass over each stiffening panel so that it overlaps the edge by about 1/4 inch. Wet it down and smooth the overlap onto the triax. Trim as needed. The overlap also serves, while wet, to hold the panel in position until the bagging operation forces it into final contact with the triax. (Note that this very light glass layer serves to stabilize the panel against buckling - the critical condition. The main skin thickness is adequate for tension and "handling" loads.)

10) Place peel ply over the bonding areas for the spar, all ribs, and the trailing edge. Wet in place.

11) Place perforated release film, then bleeder over the entire work area. Place at least four vacuum valve bottoms, anywhere but on a stiffening panel.

12) Attach the bag, connect the vacuum system, and pump down, as previously described. Inspect the setup for any shifting of the stiffening panels.

13) Allow to cure thoroughly before attempting to remove from the mold.

Tutorial Test Sections

The above procedures are intended only to suggest techniques. As with other technologies, including woodworking and metalworking, there is no substitute for actual experience. A craftsman learns by trial to how feed a drill press if drilling steel or wood. The metalworker learns the feel of bending metal and how to buck rivets. With composite work, the number of variables that the builder must deal with and get a "feel" for is at least as great as with these related crafts.

A builder could obtain the detailed plans for one of the several composite designs available and build the aircraft from scratch exactly as described, without deviation. (In fact, the point has been made repeatedly throughout this book that this is the only safe way, unless the builder is willing to be responsible for his own testing and design.) The result would be a very satisfactory aircraft and an excellent learning experience, but would not necessarily demonstrate the full potential of modern composite design available to the amateur.

The techniques described here are easily learned by experiment, with very little investment of time or money. In fact, a vacuum pump is probably the only piece of equipment that a home shop owner would not already own.

Several very basic experiments that have great teaching value are outlined below. A few days spent trying these experiments will get a lot of mistakes out of the way.

1) Borrow a large glass mixing bowl from the kitchen and use it as a mold for a simple non-bagged wet layup. Experiment with mold release. Experience directly how satin-weave cloth can conform to a dished curve and how to handle wet prepreg. Build up four layers of 9-ounce cloth. Do the B-phase knife trim around the edge of the bowl, like trimming a pie and get a feel for when is too early for the trim (and when is too late!) Let it cure and take it out of the mold.

2) Obtain a smooth flat tool surface, like a piece of sheet metal or Kortron, and make a vacuum bagged foam sandwich layup like is outlined in Example 1. Try it with and without release film on the tool and with and without wax on the tool (but not without one!). Try the entire experiment with hardware store polyethylene, using no industrial bagging film at all. Get a feel for the action of the perforated release film and how much bleeder is required. It would be useful to try this experiment several times with a range of foam types and foam thicknesses. Try making a flat layup covering most of an entire sheet of Kortron. Include

several different foam stiffening plates of various foam thicknesses and skin thicknesses. Get a feel for the limits of working to marks made on release film where the marks must be visible through the wet layup.

3) Hot wire a piece of stryrofoam to make a trial plug of convenient size, maybe shaped like a wing section. Surface it with drywall joint compound, smooth, and paint with lacquer, preferably white. Make partition flanges, as described, and add reference indentations. Make a small splash mold. Experiment with mold release techniques, including use of brown plastic mailing tape as a release surface. Try to make a plug that produces a very smooth mold surface.

4) Try making a vacuum bagged part from the test mold. Include a stiffening plate in the layup, as in Example 3.

Chapter 15
Conclusion

An airplane only "knows" its shape and weight distribution. The passing air molecules don't care what the airplane is made of. A pilot could not tell by observing the performance of two otherwise identical aircraft whether they were made of wood-and-wire, metal, or composite.

The author regularly flies a Beech *Sundowner*. I had the opportunity to fly the prototype of the *KIS Cruiser*, by Tri-R Technologies, with four big men aboard - near gross. The *KIS Cruiser* is a fixed gear, fixed pitch, aircraft with the same Lycoming O-360 as the Sundowner. There were important differences of pilot impression. The *Cruiser* felt cleaner, like a *Bonanza*, and had outstanding slow flight characteristics with the heavy load. It is hard to attribute these flight characteristics only to composite construction, but the feeling of quiet tightness is probably due to the good damping of the sandwich structure. It is one of the few aircraft where the gross weight is twice the empty weight.

The primary objective of this book is to describe the general principles of composite construction and design so that the reader will understand how and why the technology can produce excellent airplanes for reasonable effort. Composite technology is obviously not the "only" way to build an airplane, but it is certainly a good way.

Several important conclusions become clear from the discussion of the previous chapters:

1) *Materials:* A wide range of composite systems are very well suited to aircraft construction and are available to the amateur builder. Using these materials, structures can be built by the amateur that perform at least as well as with previous methods and can frequently outperform them. Composite structures will differ in detail

from, say, a design optimized for aluminum, because of different characteristics of the materials. Composite parts can be designed to equal or exceed the functions of designs in other technologies, but offer the additional great advantage of inherent corrosion resistance and resistance to damage by water.

2) Fabrication: The component materials are almost always fabricated in place, using negligible forces at ordinary temperatures. There is no equivalent to the forging of metals with great force and the welding and casting of liquid metals. Temperature-dependent functions (oven cure or post cure) are performed *after* fabrication. The processes produce strong materials which are formed before they become strong, therefore tools need not handle large forces, so are not expensive.

3) Piece Count: Composite methods inherently can produce large, complex parts with very low assembled parts count and very few fasteners. A large composite layup, like a large section of a fuselage, will require that pieces of fabric be cut to measure and put in place. The corresponding process in a metal assembly may require more small pieces of formed aluminum. Bonding with adhesives is inherently simpler than bonding with rivets.

4) Structures: Composite technology alone allows the designer to use sandwich panels to stabilize large areas against buckling so that they will transfer compression loads. This principle leads to very simple structures with large spanned areas. Where the designer in metal uses several stringers and/or hat sections to stabilize a panel, the composite designer can use a single sandwich panel. (Fabric covered designs cannot carry compression in the skin.)

5) Organization: Composite methods allow the convenient separation of the aircraft into large subassemblies for fabrication and assembly. If a fuselage is designed to be divided into a very few major sections, it is easier to assemble interior hardware like controls, instruments, wiring, and seat attachments, before the final assembly of the fuselage.

6) Replication: Some of the best applications of composite technology require that molds be made before the part can be made. The total work involved to make a mold, then to make the part is often quite acceptable, because the molds are usually simple composite splash molds themselves. A secondary benefit is that once a mold exists, it is relatively easy to make many parts.

The Kit Manufacturers

The established composite kit manufacturers offer some very important advantages to the builder/pilot. Most important: they have done the development work on both the airplane and the process and have saved the builder a very great deal of testing.

If the reader of this book were to tour the facilities of one of the kit manufacturers, he would recognize most of the processes described here, except that they would be extended to higher performance materials. Where we have discussed the use of foam panel stiffeners, the modern manufacturer may use flexible honeycomb core. With amateur methods, we wet out cloth between two sheets of polyethylene to make wet "prepreg", trim it, and put it in place. If the part cannot be vacuum bagged, we accept the weight penalty of excess resin. The manufacturer cuts out a piece of very accurately impregnated commercial prepreg, strips off the shipping film, and puts it in place on the layup. There is no excess resin and the oven cure allows the designer to reach better material performance. For similar parts, the amateur-built part will weigh more for the same function.

Considering how much work is involved in building an airplane, it is the opinion of the author that the composite kit manufacturers offer the best return for building time and money spent of any alternative. With the methods described here, the amateur could approximately duplicate the current molded kit designs, but he would have to retrace the development process of the manufacturer, without the resources, know-how, and special materials of the manufacturer. Assuming that the builder now wants a good composite airplane, the easiest and quickest way to get one is to use the design, the tooling and the skills of the manufacturer.

Potential for Designers

The Experimental Aircraft Association has many members who really do want to develop their own aircraft designs. The methods described here offer great potential for new designs.

Even if a potential designer has actually built a composite aircraft designed by someone else, it is very important to get a good feel for the technology. The designer should experiment extensively with test samples, as suggested in Chapter 14. If the designer has assembled a manufactured kit, using premolded parts, she will not have lived through much of the plug-making, mold-making, and bagging functions described. (Which is an important reason for buying the kit!) As a designer, she will have to specify these things and will need a good feel for achievable clearances, mold release problems, choices of vacuum bagging variables, etc. Where the *Dragonfly* used a solid-foam core aileron, for example, the designer might test a new sandwich skin design with 3-ounce or 1.5 ounce skins.

Large composite structures, especially with materials of mixed modulus of elasticity, can only be modeled approximately for an engineering design. A theoretical analysis is important to assist the designer in making basic design choices, but every completed design must be tested thoroughly. Martin Hollman in his *Composite Aircraft Design* (1992) and Andrew Marshall in *Composite Basics* (1993) address the problems of testing structures. It is important to test materials at the basic level with testing machines and the results are vital input for engineering calculations to design useful structures.

Critically loaded components like wing attachments and landing gear attachments should be built and tested to destruction. Very good ideas can be found by touring busy airports which have a lot of builder activity and seeing how other designers did things. Overdesign means overweight and often more work for the builder. Put the material where the load is and nowhere else.

A *Speculation*

The above discussion leads to a most interesting speculation. In principle, a builder/designer could go to an aeronautical junkyard, obtain a hopelessly corroded unflyable *Bonanza* (say), bring it home, and make a set of accurate molds from it! The builder could then clean up the impressions of all the rivet heads and smooth out some of the abrupt transitions. He might calculate the loads and forces, design the necessary internal structure, and reproduce the airplane in modern skin-stiffened, post-cured, composite, probably with a reduction in weight! And what if he used those molds to make many other copies?

Appendix 1 References

General Theory of Materials and Structures:

Gordon, J.E. *The New Science of Strong Materials*, Second Edition. Princeton, New Jersey: Princeton University Press, 1976.

> An excellent review of the principles of the theory of strong materials, with many anecdotes and examples.

(Various). *Special Issue on Materials*, September, 1967. New York: *Scientific American* magazine, 1967

> Outstanding summary of the basic properties of advanced materials. Old, but the principles are still valid.

Theory of Composites:

Clauser, H. R. *Advanced Composite Materials* New York: *Scientific American* magazine, July, 1973

> An early, but clear, description of the internal structure of high performance composites.

Kelly, Anthony. *Fiber-Reinforced Metals* New York: *Scientific American* magazine, February, 1965

> Mechanics of failure

Slayter, Games. *Two-Phase Materials*. New York: *Scientific American* magazine, January, 1962

> One of the first popular descriptions of the principles of fiber composite materials.

Composite Techniques

Hollman, Martin. *Composite Aircraft Design*, 3rd Edition
Monterey, California: Aircraft Designs, Inc., 1991

> A summary of available aircraft composite materials and analytical design techniques. The mathematical examples are probably beyond most readers of this book. Good description of testing equipment and procedures.

Hollman, Martin. *How To Build Composite Aircraft*
Monterey, California: Aircraft Designs, Inc., 1993

> A brief summary of the procedures used to manufacture and assemble modern composite kits. Good description of vacuum bagging with hardware and commercial bagging supplies specifically named. Good description of procedures for building a Rutan-type moldless fuselage. Photographs showing assembly of the Stallion and the Lancair IV.

Kress, Gregory. *Advanced Composites*
(Unpublished Syllabus for Builder's Workshop).
Griffin, Georgia: Alexander SportAir Centers, 1996

> Outline, brief background material, and shop notes for EAA/Aircraft Spruce weekend workshop for builders. Exercise on lamination and vacuum bagging.

(—). *Composite Basics*
(Unpublished Syllabus for Builder's Workshop).
Griffin, Georgia: Alexander SportAir Centers, 1996

> Very good brief summary of basic moldless materials and techniques. Used at EAA/Aircraft Spruce weekend workshop for builders.

Marshall, Andrew C. *Composite Basics,* 3rd edition
Walnut Creek, California: Marshall Consulting, 1993

> Good summary of the nature of current composite materials. Gives examples of the design and testing of typical fittings, using composite layup. Good treatment of cores and core materials.

Lambie, Jack. *Composite Construction for Homebuilt Aircraft* Hummelstown, Pennsylvania: Aviation Publishers, 1984

> Good summary of aerodynamic principles as applied to small aircraft. Overview of methods with reference to specific designs.

Loken, H. and Hollman, Martin. *Designing With Core*
Monterey, California: Aircraft Designs, Inc., 1988

> Essentially a companion to other Hollman books listed. Description of fabrication techniques, design and analysis, including Finite Element Analysis. Directed mostly toward oven-cured industrial honeycomb cores as used by kit manufacturers.

Rutan, Burt. *Moldless Composite Homebuilt Sandwich Aircraft Construction,* 2nd Edition.
Mojave, California: Rutan Aircraft Factory, 1980

> Detailed description of procedures used to build the LongEze. Classic. Still valuable.

(—). *Epoxy Plastic Tooling Manual,* (Rev. 4/90)
East Lansing, Michigan. CIBA-GEIGY Corp. Formulated Systems Group

> Excellent illustrated procedures manual for aircraft tooling. Directed toward the main-line aircraft industry, so some materials and methods do not apply, but very clear on methods.